Surviving the Desert

Greg Davenport's Books for the Wilderness

Gregory J. Davenport

STACKPOLE
BOOKS

Copyright © 2004 by Gregory J. Davenport

For more information about Greg Davenport and Simply Survival, visit the website www.simplysurvival.com.

Published by
STACKPOLE BOOKS
5067 Ritter Road
Mechanicsburg, PA 17055
www.stackpolebooks.com

All rights reserved, including the right to reproduce this book or portions thereof in any form or by any means, electronic or mechanical, including photocopying, recording, or by any information storage and retrieval system, without permission in writing from the publisher. All inquiries should be addressed to Stackpole Books, 5067 Ritter Road, Mechanicsburg, Pennsylvania 17055.

Printed in the United States

First edition

10 9 8 7 6 5 4 3 2 1

Cover photograph by Rick Sexton
Cover design by Caroline Stover
Illustrations by Steven Davenport and Ken Davenport

Library of Congress Cataloging-in-Publication Data
Davenport, Gregory J.
   Surviving the desert / Gregory J. Davenport.— 1st ed.
      p. cm. — (Greg Davenport's books for the wilderness)
   Includes index.
   ISBN 0-8117-3071-9 (alk. paper)
   1. Desert survival—Handbooks, manuals, etc. I. Title.
GV200.5 .D375 2004
613.6'9—dc22
                                                  2003015867

To my good friend and fellow instructor Bill Frye

Contents

1. **INTRODUCTION**   1
   - The three-step approach to wilderness survival   1
   - Before you go   3

2. **DESERT CLIMATES**   5
   - Types of Deserts and How They Form   5
   - Desert characteristics   8

3. **GEAR**   10
   - Backpack   10
   - CamelBak   12
   - Tent   14
   - Bivouac bag   15
   - Poncho or tarp   15
   - Emergency all-weather blanket   15
   - Anchors   16
   - Sleeping bag   17
   - Fleece or quilted blanket   19
   - Sleeping pad   20
   - Knives   20
   - Saw   23
   - Backpacking stove   24
   - Headlamp   25
   - Cooking pots   25
   - Survival tips   25

vi   Contents

4. CLOTHING   **27**
    Understanding UV, SPF, and UPF   27
    Heat gain and loss   28
    What to wear   31
    Materials   32
    How to wear and care for your clothing   34
    Shirts and pants   36
    Parka and rain pants   38
    Boots   39
    Socks   40
    Gloves   40
    Headgear   40
    Eye protection   42
    Skin protection   42
    Survival tips   43

5. CAMPING   **44**
    Selecting a campsite   44
    Tent or bivouac bag   45
    Emergency tarp shelters   45
    Natural shelters   46
    Survival tips   57

6. FIRE   **58**
    Man-made heat sources   58
    Building a fire   59
    Fire reflector   83
    Maintaining a heat source   84
    Survival tips   85

7. SIGNALING   **86**
    Rules of signaling   86
    Signals that attract rescue   87
    Signals that pinpoint your location   88
    Cellular phones   94

	Contents	vii

	Improvised signals	94
	Helicopter rescue	97
	Survival tips	97
8.	**WATER**	**98**
	Dispelling myths about water	99
	Water indicators	100
	Natural water sources	101
	Man-made water sources	108
	Water filtration	110
	Water impurities	112
	Purifying water	113
	Survival tips	115
9.	**FOOD**	**117**
	Foods to take	117
	Plants	118
	Bugs	132
	Crustaceans	133
	Mollusks	134
	Reptiles	135
	Fish	136
	Birds	143
	Mammals	145
	Cooking methods	164
	Food preservation	167
	Food cache	170
	Survival tips	170
10.	**NAVIGATING**	**171**
	Map and compass	171
	Determining direction using the sun	190
	Determining direction using the stars	194
	Global Positioning System (GPS)	198
	Survival tips	198

Contents

11.	**TRAVELING IN HOT CLIMATES**	**199**
	How to carry a pack	199
	Basic travel techniques	200
	Terrain issues	202
	Hazards	204
	Car travel	204
	Survival tips	206
12.	**HEALTH ISSUES**	**207**
	General health issues	207
	Traumatic injuries	208
	Environmental injuries and illnesses	215
	Survival stress	221
	Survival tips	221
13.	**DESERT CREATURES**	**222**
	Snake, lizard, and animal bites	222
	Insects, centipedes, spiders, and scorpions	225
	Survival tips	230
	SURVIVAL AND FIRST-AID KITS	**231**
	CORDAGE, KNOTS, AND LASHES	**233**
	TRIP PLANNING	**241**
	INDEX	**244**

1
Introduction

Wilderness survival has many variables that dictate a person's success or failure. Each environment presents a myriad of unique challenges. Regardless of the environment, however, the same basic principles apply, from cold-weather to desert situations. Wilderness survival is a logical process, and using the following three-step approach to global wilderness survival will help you keep a clear head and proceed with meeting your needs—even under the most adverse conditions. This process is the key to survival in any environment. The only thing that differs is the order in which you meet your needs and the methods you use to meet them.

THE THREE-STEP APPROACH TO WILDERNESS SURVIVAL

1. Stop and recognize the situation for what it is, keeping a clear head and thinking logically.

    Often, when people realize they are in a legitimate survival situation, they panic and begin to wander aimlessly. This makes it harder for search-and-rescue teams to find them, and valuable time is lost that they could have spent meeting their needs. If you stop and deal with the situation—evaluating it and taking appropriate steps—your odds of survival are greatly increased.

2. Identify your five survival essentials, and prioritize them in order of importance for the environment that you are in.
    1. Personal protection (clothing, shelter, fire).
    2. Signaling (man-made and improvised).
    3. Sustenance (identifying and procuring water and food).
    4. Travel (with and without a map and compass).
    5. Health (mental, traumatic, and environmental injuries).

2   Surviving the Desert

The exact order and methods of meeting these needs will depend on the environment you are in. Regardless of the order or method you choose, these needs must be met. Various methods of meeting these needs in a desert environment are covered throughout this text.

3. Improvise to meet your needs, using both man-made and natural resources.

Once you've identified your five survival essentials and prioritized them, you can begin to improvise to meet those needs. Sometimes the answer is straightforward, and sometimes it isn't. If you need some help deciding how to best meet one of your needs, use the five steps of improvising approach:

1. Determine your need.
2. Inventory your available man-made and natural materials.
3. Consider the different options of how you might meet your need.

Greg Davenport's three-step approach to global survival

Introduction   3

Before you go, take the time to prepare.

4. Pick the one that best utilizes your time, energy, and materials.
5. Proceed with the plan, ensuring that the final product is safe and durable.

Being able to improvise is the key to a comfortable wilderness visit versus an ordeal that pushes the limits of mortality. The only limiting factor is your imagination!

BEFORE YOU GO

Before departing on any trip, take the time to review the region's climate, the gear you'll take, and your skills. Do you have the necessary skills to meet the challenge? Do you have adequate safety gear? Is the area safe for travel? If you are uncertain about the answers to these questions, it may be best to stay home. If you are traveling with a team, evaluate the other members' skills. Will they help or hinder you should something go wrong? Take the time to review emergency procedures with the whole team, and make sure everyone understands them thoroughly before you depart. Areas you should cover include, but are not limited to, the following:

- Emergency gear location (first-aid and signaling gear)
- How to operate signals and other emergency equipment
- Where you are going and emergency headings to quickly exit the area to safety

To avoid unnecessary rescue efforts, let someone know your route of travel and intended destinations, and if able, check in with them during and after the trip.

2
Desert Climates

Deserts present a survivor with a myriad of problems, including water shortages, intense heat, wide temperature ranges, sparse vegetation, sandstorms, and surface soil that is potentially irritating to the skin. There are some twenty deserts around the world, covering about 15 percent of the total land surface. Understanding the various types of deserts will help you overcome the multitude of problems each can present.

Most deserts get less than 10 inches (25 centimeters) of rainfall a year and/or have a very high rate of evaporation. What little rain there is does not come throughout the year, but usually occurs in big bursts and at irregular intervals. In some instances, dry intervals extend over several years. The desert surface is often so dry that, even during hard downpours, the water runs off and evaporates before soaking into the ground.

Most deserts lie in high-pressure zones where limited cloud cover makes the earth's surface vulnerable to the sun's radiation. As a result of constant sun exposure, the area heats up quickly, creating high temperatures. These high temperatures cause surface water to evaporate quickly. In areas with strong winds, the rate of evaporation is greatly increased.

TYPES OF DESERTS AND HOW THEY FORM

Deserts are classified by their location and weather pattern. There are high-pressure deserts, rain-shadow deserts, continental deserts, and cool coastal deserts.

HIGH-PRESSURE DESERTS

High-pressure deserts occur at the polar regions and between 20 and 30 degrees latitude on both sides of the equator. These deserts are located in areas of high atmospheric pressure where ongoing weather patterns cause dry air to descend. As the dry air descends, it warms up and absorbs much of the moisture in the area.

High-pressure deserts in the polar regions
People often don't think of the polar regions as having deserts because of the cold temperatures. But there are polar areas with an annual precipitation of less than 10 inches a year that qualify as deserts. A polar desert rarely has temperatures over 50 degrees F and often has day and night temperature changes that cross over the freezing point of water.

High-pressure deserts between 20 and 30 degrees latitude
High-pressure deserts located between 20 and 30 degrees latitude north or south of the equator are hot as a result of the wind's weather pattern and their proximity to the equator. These deserts have been known to reach temperatures as high as 130 degrees F. Most of the world's deserts, including the Arabian Desert and the Sahara Desert, are located in this area.

RAIN-SHADOW DESERTS
Rain-shadow deserts occur as a result of a mountain range's effects on the prevailing winds. As wind travels over a mountain range, it cools and dumps its moisture in the form of rain or snow. As it descends to lower elevations on the other side of the mountain range, the wind becomes very dry and warm. Unless moisture is provided in some other form, a rain-shadow desert will form on the protected side of the mountain range as a result. Rain-shadow deserts include the Patagonian Desert, created by the Andes, and the Great Basin Desert, created by the Cascade Range.

CONTINENTAL DESERTS
Continental deserts occur in the centers of large continents. As inland winds travel from the sea over land, they lose moisture in the form of rain, and by the time they reach the center of a large continent, they are very dry. Continental deserts include portions of the Australian Desert and the Gobi Desert.

COOL COASTAL DESERTS
Cool coastal deserts are the result of the cold ocean currents that parallel the western coastline near the Tropics of Cancer and Capricorn. At these

Desert Climates 7

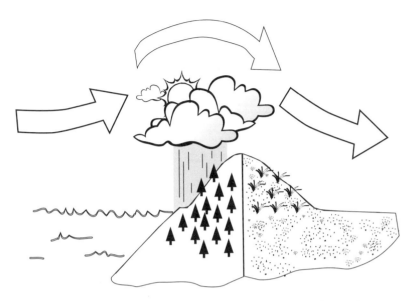

A rain-shadow desert forms when air loses its moisture as it travels over a mountain range.

Rain-shadow desert

Gobi Desert

locations, the cold ocean current touches a warm landmass, and as a result, almost no moisture is transferred from the ocean's cold water to the air that flows over the adjoining coastline. The descending air mass, which is already dry, becomes even drier. These deserts are some of the driest in the world. Cool coastal deserts include the Atacama Desert of South America and Mexico's Baja Desert.

DESERT CHARACTERISTICS

TERRAIN
Approximately 20 percent of the world's deserts are covered in sand that often resembles unmoving ocean waves. Half of all deserts are gravel plains—extensive areas of level or rolling, treeless country created by the wind's removal of ground soil, leaving only loose pebbles and cobbles. The remaining desert terrains include scattered barren mountain ranges; rocky plateaus, often seen as steep-walled canyons; and salt marshes, flat desolate areas with large salt deposits.

CLIMATE

Deserts may be both hot and cold and may or may not have seasonal rainfall. However, most deserts have large temperature swings between day and night as a result of low humidity and clear skies. In addition, desert winds increase the already prevalent dryness in the atmosphere.

VEGETATION

Little plant life is found in deserts due to the hostile environment created by the lack of water and temperature extremes. Plants that survive do so through drought escaping, rapidly reproducing when rain arrives; drought resistance, storing water in their stems and leaves; drought enduring, efficiently absorbing what little water they receive; or obtaining water from sources other than precipitation. As an adaption to the sun's unrelenting heat, many desert plants have small leaves oriented in a near vertical position. To avoid being consumed by herbivores, most desert plants have thorns, spines, and chemical compounds such as tannins and resins.

ANIMALS

A wide assortment of wildlife can be found in deserts. In order to survive, most creatures avoid the temperature extremes. Most small game animals live in burrows during the day and come out at night, and some remain dormant during the rainless seasons. Larger game animals are often active during the day but routinely seek shade during the hottest hours. Most desert creatures have learned to compensate for the lack of water by developing the ability to meet this need from the food they metabolize.

3
Gear

When traveling into a desert environment, the type of gear you carry can either help or hamper your efforts. Take the time to choose tools that will make your travel and stay more comfortable.

BACKPACK

Two basic pack designs are used by backcountry travelers: internal- and external-frame packs.

EXTERNAL-FRAME BACKPACK

The external-frame pack uses a frame that holds the pack away from your back. This is an advantage when traveling in hot weather, but it also makes the pack prone to sudden shifts that can occur without warning and disrupt your balance. An external-frame pack is best when used in extremely hot weather (during nontechnical travel) and when hiking on a trail.

INTERNAL-FRAME BACKPACK

When hiking off-trail, an internal-frame pack is preferable. This pack rides low on the body and close to the back, which allows you better balance as you travel. In hot climates, look for a frame that provides some degree of ventilation between your back and the pack. To do this, some manufacturers use a synthetic open mesh or ridges on the back of the pack that help wick moisture away from the body and allow for better air circulation.

The size of the pack you'll need depends on its use. For overnight trips, 3,000 to 5,000 cubic inches are sufficient. For long trips, you'll need 5,000 or more cubic inches. The pack should fit your back's length and contour, have strong webbing, provide thick shoulder and waist padding, and have external pouches to carry extra water bottles.

Gear 11

An external-frame pack allows better circulation between the pack and your back.

To make an improvised backpack, start by finding a forked branch (sapling or bough), and cut it 1 foot below the fork and 3 feet above. Trim off excess twigs, cut notches about 1 inch from each of the three ends, and tie rope or line around the notches of the two forked branches. Bring the two lines down and tie them to the notch on the single end of the sapling to create the pack's shoulder straps. Make sure your shoulders can fit

12   Surviving the Desert

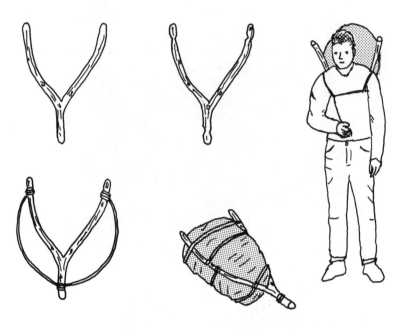

Improvised backpack

through these loops and that the line is not too tight or too loose. Place your gear inside a waterproof bag and attach it to the forked branch. To carry this pack, create a chest strap that runs through the shoulder straps at armpit height. This line should be long enough that you can hold its free end in your hand to control the amount of pressure exerted by the pack on your armpits and shoulders.

CAMELBAK

The CamelBak is a water carrier with drinking nozzle that provides easy access to drinking water. It not only ensures that you'll stay hydrated, but also gives you a way to carry more emergency gear in its cargo pockets. When in camp, I always have it on. On the trail, it is secured to the top of my large pack with the water bladder's hose draped over a shoulder strap. This gives me continual easy access to my water, and I can quickly get into my emergency gear when I stop. I have several sizes, all of which allow me to carry 100 ounces of water. The HAWG can carry 1,203 cubic inches

The CamelBak is a great way to carry your water and emergency gear.

(1,020 in the cargo pocket) and measures 9 by 7 by 19 inches. It weighs 1.9 pounds empty and approximately 8.2 pounds when its 100-ounce water reservoir is filled. The HAWG CamelBak costs about $100.

TENT

The majority of tents are made of nylon and held up with aluminum poles. In deserts, a tent provides shade during the day and warmth at night and helps protect you from most creatures. The ideal desert tent will have lightweight and durable poles and provide optimal ventilation through micromesh walls and dual-sided doors that can be rolled up. In addition, the tent should come with a full coverage rain fly that can be used when you need protection from harmful ultraviolet (UV) rays (see chapter 4 for details on UV effects).

A tent's size, strength, and weight will all factor into your decision on which one to use. A balance—or trade off—between weight and strength is often foremost in people's minds. When choosing a tent, you'll have to decide which is more important, less weight on your back or more durability and comfort in camp. Tents that are dark colored will absorb more heat and should be avoided unless you are traveling in a cool desert area. Since moisture from inside the tent will escape through the breathable wall and collect on the inside of the rain fly, make sure the two walls do not touch. If they do touch, the moisture will not escape and condensation will form inside the tent. The ideal rain fly will allow for a small area of protection between the door and outside, commonly called a vestibule. This area allows for extra storage, boot removal, and cooking. As with clothes, a zipper with a dual separating system (separates at both ends) and teeth made from a material like polyester is advised. To waterproof these zippers, make sure that either a baffle covering is used or a waterproof coating is applied to the zipper's backing. The latter example has the benefit of decreased weight and easier access to the zipper.

Most tents are classified as either three-season or four-season tents. Combo tents combine features of both. In desert conditions, a lightweight three-season tent is your best option.

THREE-SEASON TENTS

Three-season tents are typically lighter and often have see-through mesh panels that provide ventilation, which is ideal in hot environments.

FOUR-SEASON TENTS
Four-season tents are made from solid panels and in general are heavier and stronger. Typically they have stronger poles and reinforced seams, and are ideal in many cold environments.

COMBO TENTS
Some tents are marketed for either three or four seasons by providing solid panels that can be zipped shut over the ventilating mesh.

BIVOUAC BAG
Though the original concept of the bivouac bag was to provide the backpacker an emergency lightweight shelter, many travelers now carry them as their primary three-season shelter, even though they are made for just one person. Good bags are made from a breathable waterproof fabric such as Gore-Tex or Tetra-Tex, with a coated nylon floor. A hoop or flexible wire sewn across the head area of the upper surface, along with nonremovable mosquito netting, is advised for comfort and venting when needed. As with tents, the bivouac bag should have a dual separating zipper with teeth made from a material like polyester, and either baffles or an applied waterproof coating.

PONCHO OR TARP
A poncho or tarp is a multiuse item that can meet shelter, clothing, signaling, and water procurement needs. Its uses are unlimited, and you should take one along on most outdoor activities. The military ripstop nylon poncho measures $54\frac{1}{2}$ by 60 inches and features a drawstring hood, snap sides, and corner grommets.

EMERGENCY ALL-WEATHER BLANKET
Don't waste your money or risk your life carrying one of those flimsy foil emergency blankets. Instead, carry a durable, waterproof, 10-ounce, 5 by 6-foot all-weather blanket. These blankets are made from a four-ply laminate of clear polyethylene film, a precise vacuum deposition of pure aluminum, a special reinforcing fabric (Astrolar), and a layer of colored polyethylene film. The blanket will reflect and help retain more than 80 percent of the body's radiant heat. In addition to covering the body, these blankets have a hood and inside hand pockets that aid in maintaining your body heat.

16   Surviving the Desert

Emergency all-weather blanket

When compressed, these blankets take up about twice as much space as the smaller foil design, but the benefits far outweigh the size issue. The all-weather blanket is a multiuse item that can double as an emergency sleeping bag, signal, poncho, or shelter.

ANCHORS
In desert regions, it may be difficult to drive a stake into the ground or find underbrush necessary to put up a shelter or improvise items to use in your camp. To compensate for these problems, carry along several pieces of 2-foot-square ripstop material and para cord. To make an anchor, attach three or four pieces of 2-foot line to a piece of the material in various locations, and tie the lines' free ends together with an overhand knot. Attach another line to the overhand knot, and run it to your shelter. Fill the material

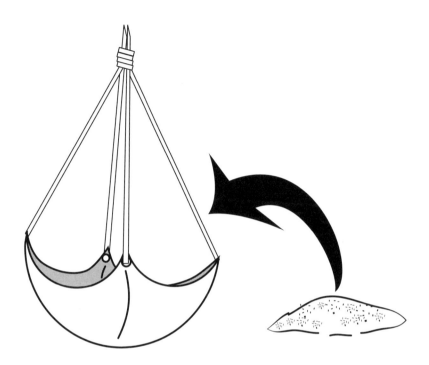

An improvised anchor helps you secure a shelter in place.

with sand, and place it so that it holds the shelter in place at the desired location. Use as many anchors as needed to hold the shelter securely.

SLEEPING BAG
Many types of sleeping bags are available, and the type used varies greatly among individuals. There are several basic guidelines you should use when selecting a bag. The ideal bag should be compressible, have an insulated hood, and be lightweight but still keep you warm. Most manufacturers rate their bags as summer, three-season, or winter expedition use and provide a minimum temperature at which the bag will keep you warm. This gross rating is used as a guide and should help you select the bag that best meets your needs. Sleeping bag covers are useful for keeping your bag clean and add an extra layer of insulating air. How well the bag keeps you warm

depends on the design, amount and type of insulation and loft, and method of construction. When selecting a bag, don't forget that deserts can get very cold at night. I'd advise against bags that have a minimum temperature rating higher than 20 degrees F.

DESIGN

Without question, the hooded, tapered mummy style is the bag of choice for all conditions. In cold conditions, the hood can be tightened around your face, leaving a hole big enough for you to breathe through. In warmer conditions, you may elect to leave your head out and use the hooded area as a pillow. The foot of the bag should be somewhat circular and well insulated. Side zippers need good, insulated baffles behind them.

INSULATION

Sleeping bags will use either a down or synthetic insulation material.

Down

Down is lightweight, effective, and compressible. A down bag is rated by its fill power in cubic inches per ounce. A rating of 550 is standard, with values increasing over 800. The higher ratings provide greater loft, meaning a warmer bag. The greatest downfall to this insulation is its inability to maintain its loft and insulating value when wet. In a desert climate, this may not be an issue, so down bags are an excellent, albeit expensive, option.

Synthetic insulation

Synthetic materials provide a good alternative to down. Their greatest strength is the ability to maintain most of their loft and insulation when wet along with the ability to dry relatively quickly. On the flip side, they are heavier and don't compress as well as down. Although cheaper than down, they tend to lose loft more quickly over long-term use. Lite loft and Polarguard are two great examples of synthetic insulation. If you are traveling in a desert area during its rain season, it is best to carry this type of bag.

Method of construction

Insulation material is normally contained in baffles, tubes created within the bag. There are three basic construction designs for sleeping bags: slant

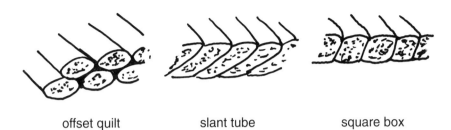

offset quilt   slant tube   square box

Sleeping bag construction designs

tube, offset quilt, and square box. Each design has its benefits, and the type you choose depends on many factors, including weight, temperature, and compressibility. However, the slant tube and offset quilt are more comfortable and provide better insulation from the ground.

IMPROVISING A BAG
Understanding the basic bag design is the key to improvising a bag in a time of crisis. I once made a bag using ripstop parachute material, dry leaves, and moss.

FLEECE OR QUILTED BLANKET
A blanket can be used to increase the insulating ability of your lightweight sleeping bag or carried as emergency gear when taking day hikes in the desert.

FLEECE BLANKET
A fleece blanket is soft, comfortable, and durable, and some say it provides more warmth for its weight than wool. The biggest drawbacks are its inability to repel wind and its lack of compressibility.

ENHANCED INFANTRY THINSULATE PONCHO LINER
This lightweight quilted blanket measures 91 by 60 inches and can be used in emergencies and for general survival needs. The newer lightweight Thinsulate blanket provides superior warmth and can be compressed down 25 percent smaller than its predecessor, which was made from polyester.

SLEEPING PAD

A sleeping pad is essential for insulating you from the ground. Most commercial pads are closed-cell or open-cell foam or a combination of the two. Each style has its pros and cons.

CLOSED-CELL FOAM

These pads provide excellent insulation and durability but are bulky to carry. They may or may not have an outer nylon shell covering.

OPEN-CELL FOAM

These pads are often self-inflating, using a high-flow inflation valve. Their ability to compress and rebound makes them ideal when space is a concern. Open-cell foam pads are usually covered with a durable, low-slip polyester fabric.

IMPROVISING A PAD

If you don't have a pad, you can improvise one using boughs, moss, leaves, or similar dry materials. Make an 18-inch-high mound that is large enough to protect your whole body from the ground.

KNIVES

You'll need a folding pocketknife for the majority of your cutting work and a larger, fixed-blade knife or saw for the bigger projects. I consider the pocketknife one of my most important tools and use it in virtually all of my improvised tasks, including cutting line, improvising shelter, preparing fire, and skinning game. The weakest part of a folding-blade knife is its lock—the part that keeps the blade open and prevents it from closing on your fingers. A good lock will secure the blade tightly to the handle when it is open. I prefer a blade length of 3 inches.

For most big projects, such as cutting dead sage branches or prepping the larger stages of firewood, a large, fixed-blade knife is all you need. Avoid knives that have multiple modifications to the blade that supposedly allow you to do the unimaginable; it's just a bunch of marketing hype. I prefer a 7- to 9-inch knife blade, with the total length measuring around 15 to 17 inches.

Gear

The Benchmade knife has a patented axis lock that sets the standard for the folding-blade knife.

A knife has many uses and is probably one of the most versatile tools you can carry. The potential for injury is high, however, so take every precaution to reduce this risk. Always cut away from yourself and maintain a sharp blade.

The SCOLD acronym—Sharp, Clean, Oiled, Lanyard, and Dry—can help you remember the proper care and use of your knives.

SHARP
A sharp knife is easier to control and use, decreasing the chances of injury. Two methods of sharpening a knife are outlined below. To establish the best sharpening angle, lay the knife blade flat onto the sharpening stone, then raise the back of the blade up until the distance between it and the stone is equal to the thickness of the blade's back side. To obtain an even angle, repeat the sharpening procedure on both sides of the blade. Each side should be done the same number of times.

Bill Seigle makes solid performing fixed-blade knives.

Push-and-pull technique
In a slicing fashion, repeatedly push and pull the knife's blade across a flat sharpening stone. If a commercial sharpening stone isn't available, use a flat, gray sandstone (often found in dry riverbeds). For best results, start with the base of the blade on the long edge of the stone, and pull it across the length of the stone so that when you're done, its tip has reached the center of the stone.

Circular technique
In a circular fashion, repeatedly move the knife blade across a circular sharpening stone or gray sandstone. Starting with the base of the blade at the edge of the stone, move the knife in a circular pattern across the stone.

CLEAN
Dirt and sand that get into the folding joint can destroy it and cause it to freeze closed or open, or even break. Dirt and sand can also be harmful to the blade's steel and can lead to its deterioration. To clean a knife blade,

use a rag and wipe it from the backside to avoid cutting yourself. Never run it across your pants or shirt, which would transfer the dirt into the pores of your clothing and risk a cut. Use a twig to help get the cleaning rag into hard-to-reach spots.

OILED
Keeping the knife's blade and joint oiled will help protect them and decrease the chances of rust.

LANYARD
Before using your knife, attach it to your body with a lanyard. To determine the lanyard's proper length, hold the knife in your hand and fully extend your arm over your head, then add 6 inches. This length allows you full use of the knife and decreases the risk of cuts due to a lanyard that is too short.

DRY
Keeping your knife dry is an important part of preventing rust, which can ruin the blade and the joint.

SAW
Consider taking a Pocket Chain Saw or Sven Saw along on trips into cold deserts. Either will allow you to break down bigger sections of firewood into a more workable size. During a cold desert night, you'll be glad you brought one.

POCKET CHAIN SAW
The 31-inch heat-treated steel Pocket Chain Saw weighs only 6.2 ounces when stored inside a 2¾-by-⅞-inch tin can. The saw has 140 bidirectional cutting teeth that will cut wood just like a chain saw; the manufacturer claims it can cut a 3-inch tree limb in less than ten seconds. The kit comes with two small metal rings and plastic handles. The rings attach to the ends of the saw blades, and the handles slide into the rings to provide a grip that makes cutting easier. To save space, however, I don't carry the handles and simply insert two sturdy branches, about 6 by 1 inch, into the metal rings. The Pocket Chain Saw costs around $20.

SVEN SAW

The lightweight Sven Saw is made from an aluminum handle and a 21-inch steel blade that folds inside the handle for easy storage. When open, the saw forms a triangle measuring 24 by 20 by 14 inches; when closed, it measures 24 by 1½ by ½ inches. The saw weighs 16 ounces and costs around $22.

BACKPACKING STOVE

When selecting a stove, consider its weight; the altitude and temperatures of where you are going; the stove's ease of operation, even in cold or windy conditions; and fuel availability. The two basic styles are canister and liquid fuel. Canister designs use butane, propane, or isobutene cartridges as their fuel source. The most common types of liquid fuels used are white gas and kerosene.

BUTANE OR PROPANE

A canister allows for a no-spill fuel that is ready for immediate maximum output. Butane and propane canisters are available throughout the United States and most of the world. I like these types of stoves due to their ease of use and unmatched performance. Some versions do not perform well in temperatures below freezing, however, and disposal of the used canisters and availability of fuel may sometimes be problems.

WHITE GAS

White gas has a high heat output and is readily available in the United States. Although the fuel quickly evaporates, it is highly flammable if spilled. The stove often does not require priming in order to start.

KEROSENE

Kerosene has a high heat output and is available throughout the world. Unlike white gas, when spilled this fuel evaporates slowly and will not easily ignite. The stove requires priming in order to start.

USING A STOVE

The exact use of the stove depends upon the manufacturer's recommendations and the type of fuel you use. As a general rule, a windshield is a must, preheating the stove helps it work better, and a stove that has a pump performs better when pumped up. For safety purposes, don't use a stove in a tent or enclosed area, except when considered absolutely necessary. If you do, make sure the area is vented and do everything in your power to avoid fuel leaks. Always change canisters and lines, fill the fuel tank, and prime the stove outside of the shelter.

HEADLAMP

A headlamp has become a great alternative to the old hand-held flashlight. The greatest benefit of a headlamp is that it frees up your hands so that you can use them to meet your other needs. When selecting a headlamp, consider its comfort, battery life, durability, weight, water resistance, and whether it will have a tendency to turn on while in a pack. I prefer the newer style headlamps, which provide a compact profile with the battery pack directly behind the bulb.

COOKING POTS

Cooking pots are luxury items that can be used to cook food and boil water. Many types are available; the kind you should use depends on your needs. I recommend a cookware set that includes a frying pan that doubles as a lid, several pots, and a pot gripper or handle. Pots are available in four basic materials:

Aluminum: Aluminum is cheap and the most common material used by backpackers. Unless the pan has a nonstick coating on the inside, however, plan on eating scorched food.

Stainless steel: Stainless steel is far more rugged than aluminum but weighs considerably more.

Titanium: Titanium is lighter than aluminum, but the cost may be prohibitive. It has a tendency to blacken your food unless you constantly stir it.

Composite: Composite cooking pots are durable yet lightweight. The inside is made from steel to reduce scorching, and aluminum is used on the outside to decrease weight.

SURVIVAL TIPS

MULTIUSE ITEMS

Try to carry gear that can serve multiple uses when possible. For example, a durable all-weather blanket (not a flimsy foil one) that has an orange and a silver side not only augments your clothing, but also can be used as a signal, water collection device, or shelter. A military poncho, heavy-duty garbage bag, and parachute line are other multiuse items that you might carry into the desert.

SLEEPING WARM

It is common for deserts to have cold nights, and since sleeping bags work by trapping dead air, make sure to fluff your bag before getting inside. Exercising and eating a protein snack before bed will help your body produce the needed heat to keep you warm once inside your bag. To avoid condensation inside the bag, keep your mouth and nose uncovered. If conditions are extreme, cover your face with a T-shirt or other porous material. On hot nights, avoid overheating by opening up the bag. Don't wait until you're sweating.

TAILOR GEAR TO CLIMATE

Deserts are known to have extreme temperature swings between night and day, and although it is tempting to carry less gear into the desert, don't. The cold nights can lead to cold injuries just as much as the hot days can result in heat injuries.

4
Clothing

It is estimated that over one million Americans will develop skin cancer each year. Considering these statistics, it is hard to imagine why an individual would travel into a desert without protection from the sun. Desert hikers can receive direct sun exposure to their heads, necks, arms, and legs. Some believe their clothing will protect them and dress in long-sleeved shirts and full pants and wear broad-rimmed hats. However, unless this clothing is able to block at least 93 percent of the sun's ultraviolet (UV) rays, damage will still occur. Your clothing's ability to block out UV radiation is based on its construction (knit or woven), color, and fiber count, and whether it is wet or dry. Knit fabrics tend to provide better protection than woven; dark colors are thought to provide five times more protection than white; clothes with high fiber count have better protection; and wet clothes tend to lose their ability to block UV penetration. A tightly woven cotton shirt blocks out approximately 86 percent of the sun's harmful UV rays and even less when wet. Polyester, on the other hand, has been shown to provide two to three times more UV protection than other fabrics of equal quality.

In recent years, the clothing industry has recognized the need for UV protective fabrics and has designed clothes that protect wearers from harmful solar radiation. These clothes will become as commonplace as Gore-Tex has for wet environments. Most of these clothes are variations of nylon and work by reflecting or absorbing the UV rays. When absorbed, the UV is transmitted across the fiber and released externally as heat. SolarWeave is an example of this type of fabric and provides 97 to 99 percent UV protection.

UNDERSTANDING UV, SPF, AND UPF

Since many fabrics will list their sun protection factor (SPF) or ultraviolet protection factor (UPF) instead of their UV protection, you will need to understand how they relate to one another. SPF relates to the degree to which a sun cream or lotion provides protection for the skin against the sun. The rating describes how much longer you can stay in the sun before your skin starts to burn. In other words, a rating of SPF 10 means you can stay in the sun 10 times longer with the sunscreen on than without it. SPF rating was developed for lotion and accounts for its evaporation, but this rating shouldn't be used to establish clothing's protection value. However, it often is. For optimal results, clothes need 93 percent UV protection, which equates to an SPF rating of approximately 15. A UV rating of 95 is equivalent to an SPF of 20; a UV rating of 97 is equivalent to an SPF of 33; and a UV rating of 99 is equivalent to an SPF of 100.

Ultraviolet protection factor (UPF) details how much UV radiation passes through a garment. A fabric with a UPF rating of 30 allows $\frac{1}{30}$ of the sun's UV radiation to pass through. There are three categories of UPF protection:

Good	UPF rating between 15 and 24
Very good	UPF rating between 25 and 39
Excellent	UPF rating between 40 and 50

HEAT GAIN AND LOSS

As heat is lost or gained through radiation, conduction, convection, evaporation, and respiration, you'll need to adjust your clothing to help maintain your body's core temperature.

RADIATION

Heat transfers from the environment to your body or vice versa through the process of radiation. The greatest cause of radiant heat gain occurs from the ambient air (sunlight) and heat reflection (the sun's rays bouncing off the ground). The head, neck, and hands pose the greatest threat for heat loss due to radiation. Increased layered clothing will slow the process of heat gain but doesn't stop it from occurring.

CONDUCTION

You gain body heat when you contact any item that is hotter than you are. If the item is cooler, you lose body heat. The greatest cause of conductive heat gain occurs as a result of contact with sand and rock. During the day,

Your body absorbs heat from sunlight and the sun's rays bouncing off the ground.

this exposure should be minimized. On cold nights, however, a rock may harbor enough heat to provide an added means of staying warm.

CONVECTION

Convection is a process of heat loss from the body to the surrounding colder air—a problem during a cold desert night. Unlike radiation, heat loss would not occur due to convection if you were standing completely still and there was absolutely no wind. It's the wind and your movements that cause you to lose heat through convection. Wearing clothes in a loose and layered fashion will help trap the warm air next to your body, which in turn decreases the heat lost through convection and also insulates you from the environment.

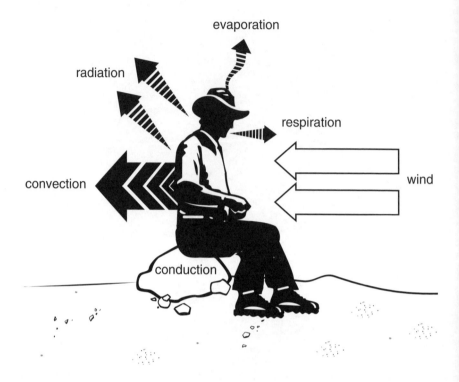

How heat is lost

EVAPORATION

Heat is lost through the evaporative process that occurs with perspiration, and since heat loss equates to calorie loss, it should be avoided as much as possible. Monitoring your activity to ensure you avoid sweating will help. If you are inactive, layered clothing will trap dead air, which will decrease the amount of heat lost through evaporation and actually keep you cooler by decreasing the amount of UV penetration.

RESPIRATION

Heat is lost through the normal process of breathing. If the night is cold, decrease the heat loss by covering or encircling your mouth with a loose cloth. By doing this, you will trap dead air and allow it to warm up slightly prior to breathing it in, decreasing the amount of heat you'll lose due to respiration.

WHAT TO WEAR

Contrary to what some might think, you should not disrobe when in the desert. One of the keys to desert survival is to protect yourself adequately from the hot sun. Loose-layered clothes protect your skin better than sunscreen and help keep you cool by trapping dead air. To both reflect heat away from the body and reduce heat loss from evaporation, wear loose and layered light-colored material that covers the entire body. Clothes for the desert should protect you from the wide temperature fluctuations and provide enough breathability to decrease the chances of overheating. Wearing a wide-brimmed hat that also covers the neck has a significant cooling effect. Sunglasses help prevent eyestrain and damage from the glaring sun. Dehydration, sunburns, and eye damage all affect your ability to meet your needs. Clothing, as your first line of personal protection, will greatly reduce your risks of injuries due to the sun.

I normally wear three layers: one that wicks moisture away, an insulating layer, and an outer shell. The ability to take a layer off or add it back when needed allows you to avoid getting too cold or overheating by adapting to the climate and the amount of work you are doing.

WICKING LAYER

Perspiration and moisture wick through this layer, keeping you dry. This is a very important layer, since having wet clothes next to the skin causes twenty-five times more heat loss than dry ones.

INSULATING LAYER

This layer traps air next to the body. Multiple layers may work better than one due to their ability to trap additional air between them. These layers help keep you cool.

SHELL

This layer protects you from wind and precipitation. The ideal shell will protect you from getting wet when exposed to rain or snow but has enough ventilation for body moisture to escape. In the desert, its greatest benefit will be how well it protects you from the wind.

MATERIALS

Clothes are made from both natural and synthetic materials. Natural materials include items like cotton, down, and wool. Synthetic materials include polyester, polypropylene, and nylon.

NATURAL MATERIALS

Cotton

Cotton has been nicknamed "death cloth," since it loses almost all of its insulating quality when wet. Cotton has extremely poor wicking qualities and takes forever to dry. Although cotton may work well during a hot desert day, it doesn't during a cold desert night. In addition, cotton's UV protection falls short of the needed 93 percent blockage advised for desert clothing. Cotton should not be worn.

Down

Down is a very good lightweight insulating material. Like cotton, however, down becomes virtually worthless when wet. Once wet, the feathers clump together and no longer trap dead air. This material is best used in dry climates or when you can guarantee it won't get wet.

Wool

Wool retains most of its insulating quality when it is wet. It also retains a lot of the moisture, however, making it extremely heavy when wet. Wool is also fairly effective at protecting you from the wind, allowing it to be worn as an outer layer. Its main drawbacks are its weight and bulkiness, making it less attractive for the desert than the lighter synthetic materials.

SYNTHETIC MATERIALS

Polyester and polypropylene

As a wicking layer, polyester and polypropylene wick well, maintain their insulating quality when wet, and dry quickly. As an insulating layer, they are lightweight and compressible. Polyester pile and fleece are commonly used for the insulating layer. They are not appropriate as an outer layer since they provide virtually no protection from the wind. A loose-fitting long-sleeved polyester shirt provides outstanding protection from the sun and is the preferred wicking layer for desert travel. If you opt to use a short-sleeved shirt, use sunscreen on your arms.

Polarguard, Hollofil, and Quallofil
Although these synthetic fibers are most often used in sleeping bags, they can also be found in heavy parkas. Polarguard is composed of sheets, Hollofil of hollow sheets, and Quallofil of hollow sheets that have holes running through the fibers. Basically, Hollofil and Quallofil took Polarguard one step further by creating more insulating dead air space. As with all synthetic fabrics listed, these materials dry quickly and retain most of their insulation quality when wet.

Thinsulate, Microloft, and Primaloft
These thin synthetic fibers create an outstanding lightweight insulation material by allowing more layers. Thinsulate is the heaviest of the three and is most often used in clothing. Microloft and Primaloft are extremely lightweight and are outstanding alternatives to the lightweight down parkas, since they retain their insulation quality when wet.

Nylon
In recent years, nylon has been used in the construction of light-colored, breathable long-sleeved shirts. In the past, nylon was used primarily as an outer layer in parkas, rain and wind garments, and mittens. Since nylon is not waterproof, most manufacturers use either special fabrication techniques or treatments to add the feature to nylon clothing used as an outer layer.

Polyurethane coatings
These inexpensive lightweight coatings protect from outside moisture, but since they are nonbreathable, they don't allow inside moisture to escape. Use this type of outer garment only when physical exertion is at a minimum.

34 Surviving the Desert

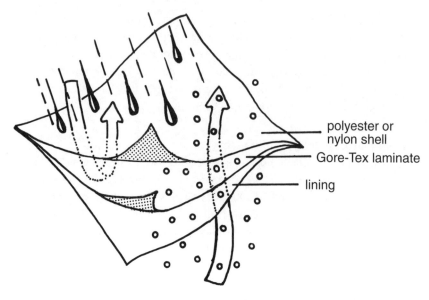

Waterproof breathable nylon

Breathable waterproof coatings
When applied to the inside of a nylon shell, this coating leaves billions of microscopic pores that are large enough for inside vapors to escape yet small enough to keep raindrops out. These coats cost more than those with polyurethane, but less than those with a breathable laminated waterproof membrane.

Breathable laminated waterproof membrane
Instead of an inner coating, a separate waterproof and breathable membrane is laminated to the inside of the nylon. The membrane is perforated with millions of microscopic pores that work under the same principle as the waterproof coating. Gore-Tex is the most common example.

HOW TO WEAR AND CARE FOR YOUR CLOTHING
For breathable fabrics to remain effective, you need to keep the pores free of dirt and sweat. Wash and dry them in accordance with the tag instructions, as some will be ruined if cleaned wrong. Don't expect the breathable coats to be perfect. If your heat output is high and you begin to sweat, this

moisture cannot escape any more than the rain can get in. To prevent this, wear your clothing in a loose and layered fashion. The COLDER acronym can help you remember how to wear and care for your clothing.

COLDER ACRONYM

C: Clean
Clothes are made of intertwined fibers that, when clean, trap dead air. As the body loses heat, it is trapped inside the dead air space and helps keep you cool or warm. The trapped air keeps you cooler when your body's radiant heat is lower than the surrounding air and keeps you warmer when it is higher. If clothes are dirty, they lose their ability to trap air and no longer insulate you.

O: Avoid overheating
If you have chosen your clothes wisely, they will allow vapor to pass through. Sweat molecules are too large to pass through and will clog the garment's dead air space, decreasing your insulation from the heat.

L: Loose and layered
Wearing multiple layers increases the amount of dead air space surrounding the body. It also allows you to add or remove individual layers of clothing as necessary for the given weather conditions. A loose-fitting, full-length garment with elbow-length or long sleeves is often worn in hot regions such as North Africa. Depending on the environment, clothes should be worn in a similar fashion and when conditions dictate in three to four layers. Suggested options are as follows:

Inner layer
This layer allows for wicking and ventilation. Polyester and polypropylene make a good inner layer. Cotton is not recommended. In hot climates, the newer lightweight polyester nylon shirts are an excellent starting layer alone or together.

Middle layer
This layer insulates by trapping dead air. Good materials for this layer are wool, polyester pile, compressed polyester fleece, Hollofil, Quallofil,

Polarguard, Thinsulate, Microloft, and Primaloft. Down can be used in dry climates or when you're sure it won't get wet. A lightweight fleece will be a welcomed layer on a cold desert night.

Outermost layer
This layer protects from wind and rain. Nylon with a polyurethane coating, waterproof coating that breathes, or laminated waterproof membrane that breathes (Gore-Tex) is best for this layer. Headgear and gloves are a must on cold nights, as one-third to one-half of body heat loss occurs from the head and hands. A lightweight Gore-Tex coat provides great protection from the harsh desert winds.

D: Dry
Wet clothes lose their insulating ability. To keep the inner layer dry, avoid sweating. Protect your outer layer from moisture either by avoiding exposure to rain or by wearing proper clothing. If your clothes do become wet, dry them by a fire or in the sun.

E: Examine
Examine clothing daily for tears and dirt.

R: Repair
Repair any rips and tears as soon as they occur. This may require a needle and thread, so make sure you pack them.

SHIRTS AND PANTS

When selecting shirts and pants for hot climates, you must consider the sun's effects. In addition to heat injuries, long-term sun exposure increases your chance of skin cancer. Wearing sun-protective clothing is one way to reduce these risks. Garments made with sun-protective fabrics have an ultraviolet protection factor (UPF) rating that designates how much ultraviolet (UV) radiation the fabric absorbs.

These ratings are based on a new garment that is dry and not worn too tight. Over time, older garments will lose some protective qualities due to repeated washings and basic wear and tear. Although darker colors provide better UV protection, a balance between this factor and how much

Clothing 37

Dressing for the desert

heat is absorbed into the fabric must be considered. Clothes that are light-colored, such as off-white, tan, or khaki, tend to absorb less heat and are probably better for this environment.

SHIRT
In hot climates, I often wear a lightweight, long-sleeved, well-ventilated nylon shirt with a UPF of 30 or more. Button tabs secure the sleeves when they're rolled up. In addition, shirts with a vented back help keep you extra cool. Nylon shirts dry quickly and provide better wind protection than polyester shirts. They are an excellent option for desert travel. These shirts can be worn alone or over a lightweight loose-fitting polyester shirt.

PANTS
Look for the same qualities in pants as in a shirt. The ideal design is made from lightweight nylon, provides UV protection, has good ventilation, and can easily convert to shorts. Cargo pockets add the ability to store emergency survival gear; any pants you buy should include this option. Pants with a drawcord at the cuff help keep critters and dirt away from your skin. Nylon pants are fast drying and provide better wind protection than most alternatives. They are an excellent option for desert travel.

PARKA AND RAIN PANTS
Although rain probably won't be a problem in the desert, wind will, so a lightweight parka and rain pants are essential. They are available in nylon with a polyurethane coating, breathable waterproof coating, or breathable laminated waterproof membrane (Gore-Tex). Some parkas come with an insulating liner that can be zipped inside. In hot climates choose a parka shell that is lightweight yet durable. Look for the following criteria when choosing a parka and rain pants:

APPROPRIATE SIZE
These garments should be big enough that you can comfortably add wicking and insulating layers underneath without compromising your movement. The parka's lower end should extend beyond your hips to keep wind and moisture away from the top of your pants.

DUAL SEPARATING ZIPPERS
Zippers should separate at both ends.

VENTILATION ADJUSTMENT
Parkas should have openings for ventilation in front, at your waist, under your arms, and at your wrist. For rain pants, the openings should be located in the front and along the outside of the lower legs, extending to about mid-calf, making it easier to put on or remove your boots. For females, pants are available with a zipper that extends down and around the crotch. The added benefit is obvious. These openings can be adjusted with zippers, Velcro, or drawstrings.

SEALED SEAMS
Seams should be taped or well bonded so that moisture will not penetrate through the clothing.

ACCESSIBLE POCKETS
What good is a pocket if you can't get to it? In addition, the openings should have protective rain baffles. Rain baffles will help keep blowing sand from damaging zippers or entering pockets.

BRIMMED HOOD
The brim will help protect you from the sun and, if it does rain, will channel moisture away from your eyes and face.

BOOTS
Sandals are not appropriate footwear for desert hikes. You need a pair of sturdy boots to provide support and protect your feet. For hot climates, lightweight leather/fabric boots are best. These are popular fair-weather boots, as they are lighter and dry faster than all-leather boots. The Danner Desert Acadia is a great boot for desert wear. It has an 8-inch upper made from leather and 1,000-denier Cordura that provides good ankle support and circulation, a liner that promotes rapid drying, and a Vibram outsole and rubber/polyurethane midsole that provide enough stability for any desert terrain. If you buy new boots, break them in before your trip.

Your boots will protect you better if you keep them clean. Wash off dirt and debris using a mild soap that won't damage leather.

SOCKS

Socks need to provide adequate insulation, reduce friction, and wick and absorb moisture away from the skin. Socks most often are made of wool, polyester, nylon, or an acrylic material. Wool tends to dry more slowly than the other materials but is still a great option. Cotton should be avoided, as it loses its insulating qualities when wet. For best results, wear two pairs of socks. The inner sock (often made of polyester or silk) wicks the moisture away from the foot; the outer sock (often a wool or synthetic blend material) provides the insulation that protects your feet. Keep your feet dry, and change your socks at least once a day. If any hot spots develop on your feet, immediately apply moleskin to prevent blisters from forming.

GLOVES

Gloves provide hand protection and decrease radiant heat loss from the hands. The type you need depends on your activity. I often take a pair of lightweight fingerless fleece gloves, which protect me during most activities and keep my hands warm during cold desert nights.

HEADGEAR

When traveling in the hot desert sun, it's essential to protect your head, face, and neck from harmful UV rays and sunburn. Headgear should be worn at all times during the day. A hat or headdress also will reduce radiant heat loss on cold desert nights. If you are working during cold nights and begin to overheat, remove your headgear only when other options, such as slowing down and adjusting your clothing layers, have not cooled you down enough. Neck-draping Sahara hats, wide-brimmed bush hats, and Arab head cloths are three great options.

SAHARA HAT

The Sahara hat has a wide forward bill that provides UPF sun protection and shades the eyes and face, as well as a rear cape that shields the neck and creates a dead air space that helps keep you cooler. The ideal Sahara hat has a moisture-wicking headband that keeps sweat out of your eyes

Clothing 41

Wide-brimmed bush hat

and a nylon strap that allows you to adjust the hat to a proper fit. This hat style is ideal for use in a desert climate.

WIDE-BRIMMED BUSH HAT

A bush hat for the desert should be lightweight and durable while providing high UPF sun protection from the sun's harmful rays. It should repel moisture, dry quickly, and have a wide brim that shades the face and neck. A moisture-wicking headband will keep sweat out of your eyes, crown grommets will help increase air circulation, and a chinstrap will keep you from losing your hat to a sudden gust of wind. These hats are often made of nylon or a similar material.

ARAB-STYLE HEADDRESS

A turban headdress usually consists of a long scarf of linen or silk wound around the head and neck. The kaffiyeh is similar but is draped over the head. To make a kaffiyeh, take a rectangular piece of cloth, fold it diagonally, and then drape it over your head. The cloth is fastened to the head by an exterior headband.

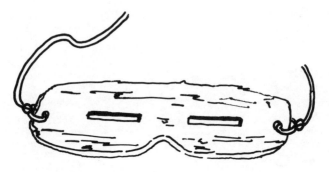

Improvised eye protection

EYE PROTECTION

Goggles or sunglasses with side shields that filter out UV wavelengths from the sunlight and reflections off bright sand are a must for travel in desert environments. It doesn't take long for the sun's reflection off the ground to burn the eyes, and once this occurs, you will have several days of eye pain along with light sensitivity, tearing, and a foreign body sensation. Since the symptoms of the burn usually don't show up for four to six hours after exposure, your eyes can get burned without your even realizing it's happening. Once a burn occurs, you need to get out of the light, remove contacts if wearing them, and cover both eyes with a sterile dressing until the light sensitivity subsides. If pain medication is available, you'll probably need to use it. Once healed, protect your eyes to prevent another burn. If no goggles or sunglasses are available, improvise by covering the eyes with either a man-made or natural material with a narrow horizontal slit cut for each eye.

SKIN PROTECTION

In a hot desert environment, ultraviolet radiation from above and reflected off the ground can be very intense and can cause painful and potentially debilitating sunburn. The best way to avoid this problem is to wear loose-fitting clothes that provide adequate UV protection. For skin that cannot be covered, use sunscreen or sunblock. Sunscreens work by absorbing the UV radiation and are available with various sun protection factor (SPF) ratings, which indicate how much longer than normal you can be exposed to UV radiation before burning. Sunblock reflects the UV radiation and is most often used for sensitive areas where intense exposure might occur, like the ears and nose. You need to constantly reapply these products throughout the day, as their effectiveness is lost over time and due to sweating.

SURVIVAL TIPS

Avoid midday sun. In hot deserts, the harmful effects of the midday sun far outweigh any benefits of working during that time. Most survival essentials can be met during dawn and dusk. At these times the temperatures are cooler and you're less apt to develop a heat-related injury. During the heat of the day, find a cool place to rest.

5
Camping

After clothing, a shelter is your next line of personal protection. When spending the night in the desert, you need a shelter to protect you from the elements: heat, cold, wind, and possibly rain. The type of shelter you use is determined by the climate, environment, available man-made and natural materials, your imagination, and your abilities as a builder. In a hot desert environment, the ground is often sand, compacted soil, rock, or gravel.

SELECTING A CAMPSITE

When selecting a campsite, make sure it provides the resources necessary to meet your needs. To find the ideal site, consider the following criteria:

LOCATION AND SIZE
Your site should be on a level surface and big enough for both you and your equipment.

DECREASE THE SUN'S IMPACT
Position the shelter so that it has a northern exposure if you're north of the equator and a southern exposure if south of it; this allows for optimal shade and decreases the sun's impact throughout the day. In addition, try to position the door so that it faces east, since an east-side opening will decrease mid- to late-day sun exposure.

AVOID WIND PROBLEMS
Since wind can wrap over the top of a tent and blow through its opening, do not place the door facing into the wind or opposite its path. Instead, have the door positioned at 90 degrees to the prevailing wind. Avoid ridgetops, box canyons, and open areas. When setting up your tent, secure it

in place by staking it down. It doesn't take much of a wind to move or destroy your shelter.

WATER SOURCE
If a water source is close by, build your shelter 100 feet or so from that location.

SAFETY FIRST
Avoid sites with the potential for environmental hazards that can wipe out all your hard work in just a matter of seconds. These include drainages or dry riverbeds with a potential for flash floods; rock formations that might collapse; dead trees that might blow down; and overhanging dead limbs. If near a body of water, stay above tide marks. Also avoid sites near large animal trails and dark, cool, damp places, which are often frequented by scorpions, snakes, centipedes, and other desert creatures.

SURVIVAL
In an emergency, locate your camp next to a signal and recovery site.

TENT OR BIVOUAC BAG
A tent or bivouac bag provides you with shade, protects you from pesky night creatures that think your sleeping bag is a nice home, and takes little time to put up. If you have a tent or bivouac bag, use it. If not, take the time to build an emergency tarp or natural shelter. For details on tent designs, refer to chapter 3.

EMERGENCY TARP SHELTERS
An emergency shade shelter can be made using your rain fly, tarp, poncho, or emergency survival blanket, should other options not be available. Although most shelter designs require a 45- to 60-degree angle on all sides, the desert shade shelter described below is made with the material nearly parallel to the ground.

If line is used, it can be attached to the shelter material's grommets or to buttons fashioned from rocks, grass, or other substance that will not tear or cut the tarp. Ball up the material inside a corner of the tarp, and tie the line around it with a slipknot. Then fasten the shelter piece to the ground using the other end of the line tied to a sandbag, vegetation, rock, big log, or improvised stake made from any available material. If using an improvised stake, pound it into the ground so it is leaning away from the tarp at a 90-degree angle to the wrinkles in the material. To avoid risking a broken hand, always hold the stake with your palm up when pounding it into the ground so that a missed strike will hit the forgiving palm versus the unforgiving back of the hand. In desert regions, it may be difficult to drive stakes into the ground, and you may need to use sand anchors (see chapter 3).

DESERT SHADE SHELTER

A desert shade shelter is most often used in dry environments. Ideally, locate an area between rocks or dunes that has an 18- to 24-inch depression. Another option is to dig an 18- to 24-inch-deep trench that is large enough for you to comfortably lie down in. Pile the removed sand around three of the four sides. To provide for an adequate entryway, remove additional sand from the remaining side. Cover the trench with your tarp or poncho, and secure it in place by weighing down its edges with sand or rocks.

If you have a second tarp or poncho, place it 12 to 18 inches above the first. To elevate the second tarp, use rocks, sand, sandbags, or any other suitable material. Layering the material increases your protection from the elements and decreases the amount of penetrating heat and UV rays. Since darker colors absorb more heat, if one tarp is darker, use it as the outer layer. The desert shade shelter will reduce midday heat by as much as 30 to 40 degrees. A second layer reduces the inside temperature even more.

To avoid sweating or dehydration, build this shelter during the morning or evening hours. Until then, get out of the heat by creating a quick lean-to. Attach a tarp to an elevated rock or sand dune, and stretch it out to form a 45- to 60-degree angle between the tarp and the ground.

NATURAL SHELTERS

In an emergency, use any available natural shelter or quickly improvise a tarp shelter. If time permits, you can build a natural shelter to provide better protection from the elements. However, the use of natural shelter materials is recommended only in an actual survival situation. As with any improvised item, a natural shelter must be constructed so that it is safe

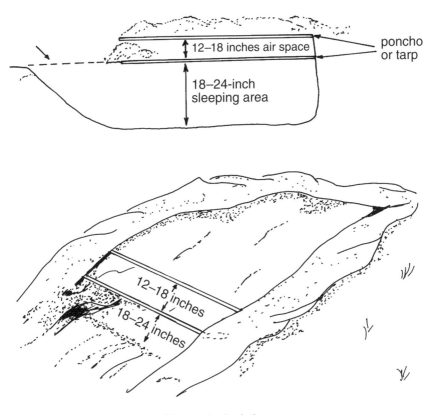

Desert shade shelter

and durable. It needs to be solid enough to withstand the elements. When using poles, they need to be strong enough to support the overall shelter. In most cases, when adequate material is available, the shelter framework is created by lashing three or four poles together and then placing additional poles on top. Use a 45- to 60-degree pitch on any roof or leaning wall. In the best-case scenario, the roof or shelter covering works best when it provides at least an 18-inch covering. All shelters should be big enough for you and your equipment. If you have a fire in or near the shelter, provide appropriate ventilation and build it in an area that has been cleared of flammable materials.

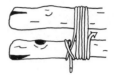 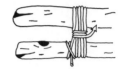 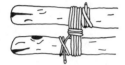

Shear lash

LASHINGS

When building your shelter, various lashings may be needed to hold the structure together. Two are the sheer lash and the square lash.

Shear lash

The sheer lash is best used when making a bipod or tripod structure. It is a simple process: Lay the poles side by side, and attach the line to one of the poles with a clove hitch. Run the line around all the poles three times in what is called a wrap. Next, run the line two times between each of the parallel poles in what is called a frap. This should go around and over the wrap. Pull it snug each time. Finish by tying another clove hitch.

Square lash

The square lash is used to join poles at right angles. Lay the poles across one another at right angles, and attach the line to one of the poles with a clove hitch. Wrap the line, in a box pattern, over and under the poles, alternating between each pole. After you have done this three times, tightly run several fraps between the two poles and over the preceding wrap. Pull it snug and finish with a clove hitch.

 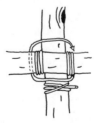 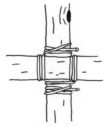

Square lash

WICKIUP

The wickiup shelter can be used anywhere that poles, boughs, brush, leaves, and grass can be found. The wickiup is not an ideal shelter during prolonged rains, but if the insulation material is heaped on thick, it will provide adequate protection from most elements.

Gather three strong 10- to 15-foot poles, and connect them together at the top with a shear lash. If any of the poles has a fork at the top, it may not be necessary to lash them together. Spread the poles out to form a tripod; a 60-degree angle is optimal. They will be able to stand without support. Fill in the sides with additional poles by leaning them against the top of the tripod. Don't discard shorter poles; they can be used in the final stages of this process. Leave a small entrance that can later be covered with man-made or improvised materials. For immediate use, cover the shelter with brush, leaves, reeds, bark, or similar materials.

For additional protection, layer on roofing materials in the following manner: Working from bottom to top, cover the framework with grass

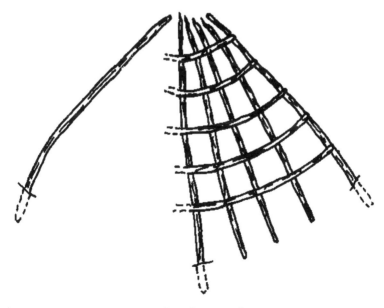

Wickiup framework

and/or plant stalks. Next, cover it, from bottom to top, with mulch and/or dirt. To hold this material in place, lay poles around and on top of the wickiup. Leave a vent hole at the top if you plan to have fires inside the shelter. The roofing options described for the wigwam may also be used to roof a wickiup.

WIGWAM

The wigwam may be a viable option in some desert regions, depending on availability of materials. The wigwam's greatest asset is the space and headroom provided by its vertical walls. This dome-type shelter provides protection from all directions and has a low wind profile. The following instructions are for a wigwam with a 10-foot floor space, providing enough room for several people and their equipment.

Wigwam framework

Cut twenty-four saplings that are 10 to 15 feet long and 2 inches in diameter. Willow and maple work best, but any sapling will do. If unable to use the saplings the day you cut them, store them so they are bent into a U shape. With a stick or your foot, mark a 10-foot circle where you intend to place the shelter. Using the circle as your guide, evenly place the saplings around it. Make holes for the saplings by pounding a solid wooden stake into the ground and then removing it. Then bury the large end of each sapling 6 to 10 inches into the ground, and tap dirt around it to help hold it in place. Next, create the basic framework by bending opposing saplings together, overlapping by at least 2 feet, and using a shear lash to secure them together. Once completed, the twenty-four poles should create a domelike structure with a center at least 7 feet high. Wrap additional saplings horizontally around the framework. Leave a 3- to 4-foot-high doorway that can later be covered with a hide or other appropriate material. For optimal roofing support, place the horizontal poles 12 to 18 inches apart.

Finally, you need to construct a roof from available natural materials. The ones most commonly used are grass (most common in deserts), mats made from various stalks, birch and elm bark, and wood shingles. Regardless of the material, a wigwam roof is constructed using a shingle design, placing the material from bottom to top, and so that the higher rows overlap those below by about one-third. Leave a vent hole at the top if you plan to have fires inside the shelter.

Grass roofing
Although grass can be used as a roofing material, its biggest drawback is the amount of time required to harvest enough to cover a shelter. Ideally, you should collect tall grass that is dry and grew the previous year. Older grass is usually brittle and rotted, and new grass must be dried before used. Separate the grass into small bundles, 1 to 2 inches in diameter, with the root ends together. Place the bundles close to one another. Fold a long piece of cordage in half, and place it about 4 inches down from the top of the first bundle (details on how to make cordage appear at the back of the book). Tie an overhand knot, slide in another bundle, and repeat. If you run out of line before you reach the end, simply tie another piece to the

Grass roofing

first and continue the process. At the same time, tie a second line about 3 or 4 inches below the first. This pattern should effectively weave the bundles together, holding them securely in place. Once you have made enough of these grass skirts to cover the shelter, lash them to the framework using proper shingling techniques as you go up the shelter.

Mat roofing

A mat made by weaving reeds or the leaves and stalks of cattails, sotol, or yucca together provides a strong covering for most shelters. Like the grass skirts, these mats are made and then attached to the shelter. Begin by laying your material down on the ground side by side. To create a tighter fit, alternate the stalks' thick and skinny ends. Fold a long piece of cordage in half, and place it about 4 inches down from the top of the first stalk. Tie an overhand knot, slide in another stalk, and repeat. If you run out of line before you reach the end of the mat, simply tie another piece to the first and continue the process. Once the first row is done, perform the same process every 4 inches down the mat until you reach the bottom. This pattern should effectively weave the bundles together, holding them securely in place.

Birch and elm bark shingles

Birch and elm are the most common types of bark used for covering a shelter. Large pieces of bark can easily be cut and stripped from a tree or log. Use a knife to make a rectangular cut from top to bottom, and peel the bark off along the vertical cut. If you experience difficulty removing the bark, beating the area with a log will help release it from the tree. Once the bark is removed, lay it flat on the ground, weight it down with a heavy material such as rocks or wood, and allow it to dry. When dry, the bark can be lashed to the horizontal beams and shingled up the shelter's framework. For best results, the shingles that are side by side should have a slight overlap. You can sew bark together if the size is not adequate for its intended use. If you don't have sewing material, you might make a needle from bone or wood or perhaps use cordage and run it through holes created with your knife.

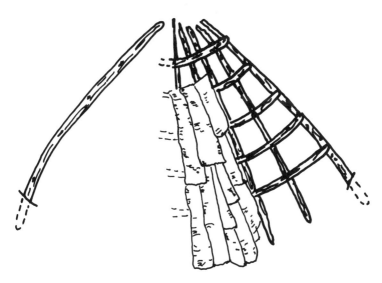

Bark shingles

Wood shingles

Straight-grained woods like cedar are the most commonly used wood roofing material. Where live cedar trees can be found, you will easily find old fallen, dead, and seasoned cedar. Without much difficulty, large half-inch-thick shingles can be removed with a knife, ax, or in some instances, your bare hands. The soft wood can be lashed or nailed to the horizontal beams and shingled up the shelter's framework. If lashed, you'll need to make holes in the top of each shingle; use an awl-like tool to perform this easy task.

A-FRAME

In some desert regions, an A-frame may be a viable option, depending on available materials. Locate a rock or dune that is 3 to 4 feet high. Place a ridgepole—any material that is 12 to 15 feet long and the diameter of your wrist—on top of the rock or dune so that a 30-degree angle is formed between the pole and the ground. Lay support poles across the ridgepole on both sides, 1 to 1½ feet apart and at a 60-degree angle to the ground. Support poles should extend above the ridgepole only slightly; if they end up above the roof material, moisture will run down them and into your shelter. Crisscross small branches, sotol or similar leaves and stalks, or cordage into the support poles (details on improvising cordage appear at

Natural A-frame

the back of the book). Cover the framework from bottom to top with any available grass, leaves, stalks, or similar material (refer to wigwams for more specifics on roofing options). Cover the door opening with your pack or similar item.

LEAN-TO

In some desert regions, a lean-to may be a viable option, depending on available materials. Find an elevated 5- to 6-foot-high, 7-foot-wide rock or sand dune, and lean multiple support poles—any material that is 12 to 15 feet long and the diameter of your wrist—against it so that a 45- to 60-degree angle is formed between the poles and the ground. Support poles need to be long enough to provide this angle yet barely extend beyond the top of the ridgepole. Weave small saplings, sotol or similar leaves and stalks, or cordage (see back of book) into and perpendicular to the support poles. Cover the framework from bottom to top with any available grass, leaves, stalk, or similar material (refer to wigwams for more specifics on roofing options). Cover the door opening with your pack or similar item. The lean-to allows you to build a fire in front of the shelter, provided it is safely placed away from you and your gear. To increase the heat's reflection into your shelter, build a fire reflector behind the fire.

CAVE

A cave is the ultimate natural shelter. With little effort, it can provide protection from the various elements. However, caves are not without risk. Some of these risks include, but are not limited to, animals, rodents, reptiles, and insects; bad air; slippery slopes, rocks, and crevasses; floods or high-water issues; and combustible gases, which are most common where there are excessive bat droppings. When using a cave as a shelter, you should follow some basic rules:

- Never light a fire inside a small cave. It may use up oxygen and may cause an explosion if there are enough bat droppings present. Fires should be lit near the cave entrance, where adequate ventilation is available.
- To avoid slipping into crevasses, getting lost, or other hazards, never venture too far into the cave.

56   Surviving the Desert

- Make sure the entrance is above the high-tide line.
- Keep a constant awareness of water movement within a cave. If it appears to be prone to flooding, look for another shelter.
- Never enter or use old mines. The risk is not worth the benefit. Collapsing passages and vertical mine shafts are just some of the potential problems.

A wall can be built at the cave entrance by leaning support poles against it and covering them with grass or mat shingles (instructions are given under wigwams). Leave an area large enough to build a fire and also to provide adequate ventilation inside the shelter.

HOBO SHELTER

To construct a hobo shelter, you need multiple pieces of driftwood and boards that have washed ashore. On the land side of a sand dune beyond the reach of high tide, dig a rectangular space that is big enough for both you and your equipment. Place the removed sand close by so that you can use it later. Gather as much driftwood and boards as you can find, and build a strong frame, using any available line, inside the rectangular dugout. Create a roof and walls by attaching driftwood and boards to the frame, leaving a doorway. If your wood supply is limited, don't place support walls at the

Hobo shelter

back or sides of the structure. This may allow some sand to fall into the shelter, but it will still be adequate. If you have a poncho or tarp that's not necessary for meeting your other needs, place it over the roof. Insulate the shelter by covering the roof with 6 to 8 inches of sand.

INCREASING INSULATION
Creating a second wall inside or out around your shelter will increase its ability to keep you warm or cool. Doing something as easy as tying mats or grass skirts to the interior walls and roof can create the second wall. To make an elaborate insulation wall, drive a row of tall stakes, about 1 foot apart, 6 to 8 inches into the ground and 12 to 18 inches from the shelter wall. Then weave willow branches or similar material between and perpendicular to the stakes. Fill in the space between walls with grass, duff, or leaves.

SURVIVAL TIPS

CHECK FOR UNWELCOME GUESTS
Before getting into a shelter or sleeping bag, check it for small creatures. They find these areas just as comfortable and inviting as you do, and they do not like to share. Odds are they will let you know this if you crawl in without letting them exit first.

6
Fire

Fire is the third line of personal protection. In most cases, it will not be necessary if you've adequately met your clothing and shelter needs. In extreme conditions, however, fire is very beneficial for warding off hypothermia and other exposure injuries during a cold desert night. Fire serves many other functions as well: providing light, warmth, and comfort; a source of heat for cooking, purifying water, and drying clothes; and a means of signaling. In addition, a fire is relaxing and helps reduce stress. For some of these purposes, building a fire is not always necessary. You might instead use a backpacking stove, Sterno stove, or solid compressed fuel tablets.

MAN-MADE HEAT SOURCES
A man-made heat source can be used in any of the various shelters, provided there is proper ventilation. If you are in a tent, however, limit its use to the vestibule area to avoid fuel spills or burning the tent.

BACKPACKING STOVE
The two basic styles of backpacking stoves are canister and liquid fuel. Canister designs use butane, propane, or isobutene cartridges as their fuel source. The most common types of liquid fuels used are white gas and kerosene. (For more details on the various styles of backpacking stoves, see chapter 3.)

STERNO
Sterno has been around for a long time and still has a place for many backcountry explorers. The fuel is a jellied alcohol that comes in a 7-ounce can.

Under normal conditions, it has a two-hour burn time. Although far inferior to a good backpacking stove for cooking, it is very effective at warming water and a shelter in an emergency. An inexpensive folding stove is made for use with Sterno, but with a little imagination you can create the same thing.

SOLID COMPRESSED FUEL TABLETS
Esbit, Trioxane, and Hexamine are the three basic compressed fuel tablets on the market. Esbit is the newest of the three, and unlike its predecessors, it is nontoxic. This nonexplosive, virtually odorless and smokeless tablet can generate up to 1400 degrees F of intense heat, providing twelve to fifteen minutes of usable burn time per cube. When used with a commercial or improvised stove, it can sometimes boil a pint of water in less than eight minutes. These tablets easily light from a spark and can also be used as tinder to start your fire.

BUILDING A FIRE
When man-made heat sources either are not available or don't meet your needs, you may elect to build a fire. Always use a safe site, and put the fire completely out, so that it is cold to the touch, before you leave. Locate the fire in close proximity to fire materials and your shelter. It should be built on flat, level ground and have adequate protection from the elements. Before starting the fire, prepare the site by clearing a 3-foot fire circle, scraping away any leaves, brush, and debris, down to bare ground if possible. To successfully build a fire, you need to have all three elements of the fire triad present—heat, oxygen, and fuel. Your fuels will vary depending on what is on hand.

HEAT
Heat is required to start a fire. The focus of this book is on wilderness survival, not primitive skills. Some deserts, however, provide abundant resources for creating an ember using a friction system, such as a bow and drill, hand drill, or pump drill. Since this could be an option, both man-made (often spark-based) and primitive (friction-based) heat sources are covered below.

Man-made and sparked-based heat sources

Man-made heat sources include matches, lighters, artificial flint, flint and steel, and pyrotechnics. Most man-made heat sources are easy to use, and at least one should be part of your emergency survival kit.

Matches

Matches run out, get wet, and never seem to work in a time of crisis. If you are dead set on using matches anyway, I recommend NATO-issue survival matches, which have hand-dipped and varnished heads that are supposed to light even when wet and exposed to strong wind or rain. These matches will burn around twelve seconds—enough time to light most fires. In order to protect the match from going out, light it between cupped hands while positioning your body to block the flame from wind or rain. Regardless of the type you carry, store your matches in a waterproof container until ready for use.

Lighter

A lighter is a form of flint and steel with an added fuel source that keeps the flame going. Like matches, it has a tendency to fail when used during inclement weather, and once the fuel is used up, it becomes dead weight. If you understand a lighter's shortcomings and still elect to use one, I recommend a Colibri Quantum. These high-end lighters are water-resistant and shockproof, ignite at high altitudes, and are marketed as wind-resistant. To use, simply place the flame directly onto the tinder.

Metal match (artificial flint)

A metal match is similar to the flints used in a cigarette lighter, but much bigger. When stroked with an object, the friction creates a long spark that can be used to light tinder. Most metal matches are made from a mixture of metals and rare earth elements. The mixture is alloyed at a high temperature and then shaped into rods of various diameters.

To use a metal match, place it in the center of your tinder, and while holding it firmly in place with one hand, use the opposite hand to strike it with your knife blade, using a firm yet controlled downward stroke at 45 to 90 degrees. The resulting spark should provide enough heat to ignite the tinder. This may take several attempts. If after five tries it has not lit,

Fire 61

Unlike matches and lighters, an artificial flint virtually never runs out.

rework the tinder to ensure that adequate edges are exposed and oxygen is able to flow within it. The S.O.S Strike Force is the most popular commercial metal match available. There also are two one-hand-use metal matches on the market: the Spark-Lite and the BlastMatch.

S.O.S. Strike Force
The S.O.S. Strike Force has a ½-inch round alloy flint attached to a hollow, hard plastic handle that houses emergency tinder. It also has a flint cover with a hardened steel striker attached, making this system completely self-sufficient. Although the system is a little bulky, it weighs slightly less than 4 ounces.

The Spark-Lite
The Spark-Lite is small and light, measuring approximately 2¼ by 9/32 by 9/32 inches. Its spark is also smaller than that of the larger metal matches.

It has a serrated wheel, similar to that of a cigarette lighter, that strikes a small flint when stroked. In order to make this a one-hand-use item, the flint is spring-loaded, maintaining contact with the wheel at all times. The small flint is supposed to allow for about a thousand strokes before it runs out. To use, stroke the sparking wheel with your thumb while holding the Spark-Lite's body with your fingers of the same hand.

BlastMatch

The BlastMatch is larger and weighs more than the Spark-Lite, measuring 4 by 1 3/8 by 7/8 inches. It has a much larger, molded plastic body that holds a 2 1/2-inch-long by 1/2-inch-diameter rod of flint. The flint is spring-loaded, and when the cap is released, the flint is propelled out. To use, place the flint tip in the center of your tinder, apply pressure to the side catch with your thumb, and push the body downward. This action will force the scraper, located inside the catch, down the flint, creating a large spark.

Flint and steel

Flint and steel are effective for starting fires, but the necessary materials may be hard to find. Some flint options are quartzite, iron pyrite, agate, or jasper. Any steel can be used with the flint, but most people use an old file. By striking the iron particles, heat is created when they are crushed and torn away.

To use the flint and steel, hold the flint in one hand and as close to the tinder as possible. With the steel in your other hand, strike downward onto the flint. Direct the resulting spark into the center of the tinder.

The best tinder to use is charred cloth, which can be created in advance by placing several 2-inch squares of cotton cloth inside a tin can with ventilation holes in its top. Place this in a fire's coals for fifteen to thirty minutes. Turn the can every couple minutes, and remove from the fire when smoke stops coming out of the holes.

Pyrotechnics

Flares should be used only as a last resort for starting a fire. It's best to save these signaling devices for their intended use. However, if you are unable to start a fire, and the risk of hypothermia is present, a flare is a

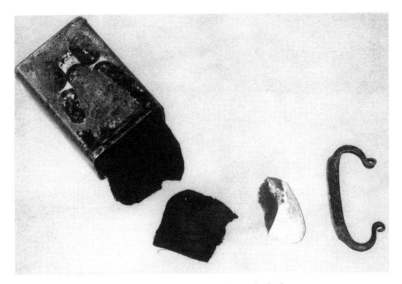

Flint and steel and charred cloth

very effective heat source. Its use is simple: After preparing the tinder, safely ignite it by lighting the flare and directing its flames onto the tinder. Time is of the essence, so prepare your firelay in advance, leaving an opening large enough that you can direct the flare's flame onto the underlying tinder.

Natural friction-based heat sources
Friction-based heat works through a process of pulverizing and heating appropriate woods, using a circular technique with a bow and drill, hand drill, pump drill, or the less-used fire plow, until an ember is created. This ember can be used to ignite awaiting tinder. The biggest problems associated with these techniques are muscle fatigue, poor wood selection, and moisture that prevents the material from reaching an appropriate temperature.

Once you have an ember, relax and take your time. Don't blow on it; the moisture from your mouth may put it out. If you feel it needs more oxygen, gently fan it with your hand. In most cases, however, simply waiting a few seconds will allow the ember to achieve its pleasant glow.

Bow and drill

The bow and drill technique is often used when the spindle and baseboard materials are not good enough to create an ember using the hand drill technique. The bow helps establish the friction needed to use materials that would otherwise be inferior or when bad weather adversely affects your ability to create an ember with the hand drill. The bow and drill is composed of four separate parts:

- *Bow.* The bow is a 3- to 4-foot branch of hardwood that is seasoned, stout, slightly curved, about ¾ inch in diameter, and has a small fork at one end. If the branch doesn't have a fork, create one by carving a notch at the appropriate place. A strong line is attached to the bow to create the tension needed to turn the spindle once it is inserted. You can use a strip of leather, parachute line, shoelace, or improvised cordage (details on how to make cordage appear at the back of the book). Securely attach the line to one end of the bow by carefully drilling a hole through the bow with a pump drill or knife, tying a knot in the line, and then running the line through the hole. The knot ensures that the line will not slip or slide forward, and since the line's tension will inevitably loosen, it allows you to make quick adjustments. Use a fixed loop to attach the line's free end to the fork on the other side.
- *Cup.* Made from hardwoods, antlers, rocks, or pitch wood, the cup has a socket for the top of the spindle. The cup's purpose is to hold the spindle in place while it is turned by the bow. When using a dead wood, you must lubricate the cup's socket to decrease the friction between the cup and the spindle. You can use body oils, animal fat, or soap shavings to accomplish this.
- *Spindle.* The spindle is a smooth, straight cylinder made from a dry, soft wood or other plant material that is approximately ¾ inch in diameter and 8 to 12 inches long. The ideal spindle is made from yucca, sotol (plants of the agave family that resemble yucca), cottonwood, aspen, willow, sage, or cactus. Dead smaller branches of cedar, locust, or ash may also be used for a spindle. The best way to evaluate the material is to press on it with a fingernail; if it makes an indentation, the material should work. To prepare the spindle for use, carve the ends so that one is cone-shaped and smooth and the other is round with rough edges.

Fire   65

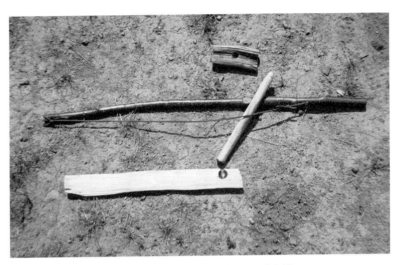

The various parts of a bow and drill

- *Fireboard.* The fireboard should be made from a material similar in hardness to the spindle. The ideal fireboard is 15 to 18 inches long, ¾ inch thick, and 2 to 3 inches wide.

Preparing the fireboard for use
1. Carve a circular socket three-quarters the diameter of the spindle, at least 4 inches from one end, close to the long side (but not right on the side), and about one-quarter the thickness of the board. If the socket is too close to the side, there will not be enough material to prevent the spindle from kicking out of the hole.
2. Prime the hole. First, twist the bowline around the spindle so that the coned end is up and the rounded blunt end is down. If it doesn't feel like it wants to twist out, then the bow's line needs to be tightened. While holding the bow and spindle together, kneel on your right knee and place your left foot on the fireboard. Insert the cone end of the spindle inside the cup, and place the round, blunt end into the fireboard socket that you created with your knife. Holding the bow in the right hand (at the closest end) and the cup in the left, apply gentle downward pressure on the spindle, keeping the spindle perpendicular to the ground. For added support and stability, rest the left arm and

elbow around and upon the left knee and shin. (If left-handed, reverse the above.) With a straightened arm, begin moving the bow back and forth with a slow, even, steady stroke. Once the friction between the spindle and the fireboard begins to create smoke, gradually increase the downward pressure and continue until a smooth, round indentation is made in the fireboard.
3. Using your knife or saw, cut a pie-shaped notch through the entire thickness of the fireboard so that its point stops slightly short of the hole's center. Place a piece of bark, leather, or other appropriate material under your fireboard for the ember to sit on. This will protect it from the moist ground and help you move it to your tinder.

Bow and drill technique

Once the separate parts of the bow and drill are prepared, it is ready to use. Simply apply the same technique used for priming the hole when preparing the fireboard. Once the smoke is billowing out and you can't continue any longer, check within or below the fireboard's notch for an ember created by the friction.

Hand drill

The hand drill is similar to the bow and drill, except that you use your hands to turn the spindle. This method is used when conditions are ideal: no moisture in the air and excellent materials. The hand drill is composed of two parts:

- *Spindle*. The spindle is a smooth, straight cylinder made from a dry, soft wood or other plant material that is approximately ½ to ¾ inch in diameter and 2 to 3 feet long. The ideal spindle is made from yucca or sotol. I've heard of people using cattail or mullein, but these materials can be finicky. To prepare the spindle for use, carve the fatter end so that it is round with rough edges. If you can't find a straight spindle, you might use a piece of cattail, bamboo, or other available reed as the shaft. Create a plug from a short piece of sotol or other material to fit inside the end, leaving 2 to 3 inches of the plug extending out of the shaft. To help prevent the end of the shaft from splitting, wrap it with sinew. You could also use such a device to drill holes by replacing the friction plug with a stone bit.

The bow and drill in use

- *Fireboard.* The fireboard is created from a soft wood or other plant material of similar but not quite the same hardness as the spindle. Yucca and sagebrush are the best desert fireboard materials. The optimal size is 15 to 18 inches long, 2 to 3 inches wide, and ½ to ¾ inch thick. Prepare a notch as described above for the bow and drill.

Hand drill technique

When using a hand drill, some people sit and others kneel. The key is to be comfortable while still able to turn the spindle and apply appropriate downward pressure. While sitting or kneeling, rub the spindle between your two hands. In order to optimize the number of revolutions the spindle

Creating a coal using a hand drill

makes, start at the top and use as much of your hands, from heel to fingertips, as you can. Apply downward pressure as your hands move down the spindle until you reach bottom, and then quickly move both hands up while ensuring that the spindle and fireboard maintain contact at all times. Since the spindle will cool rapidly, this step is crucial to your success. It's the revolutions in conjunction with the downward pressure that produces the friction needed to create an ember. When you begin to see smoke, increase your speed and downward pressure until you can't continue anymore. Just before you finish, push the top of the spindle slightly away from the fireboard's notch to help push the ember out. At times, it may be necessary to create additional downward pressure on your spindle. Two methods are commonly used:

- *Mouthpiece.* A mouthpiece is created similarly to the cup of a bow and drill, but instead of using your hands to hold it on the spindle, you use your teeth. When using this technique, shorten the spindle to 18 to 24 inches in length.
- *Thumb thong.* To make a thumb thong, tie a thumb loop at each end of a thong, and attach its center to the top of the spindle. By sliding your thumbs into the loops, you can provide a nonstop spin with increased downward pressure. As with the mouthpiece, this technique requires a shorter spindle.

Pump drill

In addition to drilling holes, a pump drill can also be adapted into a friction heat source that works similarly to the bow and drill. It consists of a spindle, crosspiece, flywheel, and fireboard.

- *Spindle.* The spindle is made from a piece of straight, debarked hardwood that measures about 30 inches long with a diameter of $1\frac{1}{8}$ inch on one end tapering to $\frac{7}{8}$ inch on the other. On the wider end, drill a 1- to 2-inch-deep hole, slightly bigger than $\frac{1}{2}$ inch diameter, into its exact center. This can be started with a hand drill with a $\frac{1}{2}$-inch stone bit (described in the Hand Drill section above), and then finished by replacing the bit with a $\frac{1}{2}$-inch wood plug of medium hardness. This technique should create a nicely rounded, burned hole that has an equal diameter throughout its depth. If the hole ends up not being centered, cut it off and start over.

Create a plug from a dry, soft wood or other plant material such as yucca, sotol, cottonwood, aspen, willow, sage, cacti, cedar, locust, or ash. This plug should fit snugly into the spindle's tip, with 2 to 3 inches extending out of the shaft. Wrap sinew around the end of the spindle for about 3 inches up its shaft to secure the plug and also help prevent the end of the spindle from splitting. On the other end of the spindle, cut a ¼-inch notch into the center of the shaft. The string from the pump drill's crosspiece will ride there. To help protect the area, wrap the end with sinew from the bottom of the notch to about 1 inch down.

- *Crosspiece.* The crosspiece is made from a piece of straight-grained, knotfree hardwood that measures 22 to 24 inches in length and about 1½ to 2 inches wide. Create a hole in the exact center of the crosspiece that is ⅛ inch bigger around than the spindle. This can be done with your hand drill or perhaps by using a hot coal. Tie a line to the far ends of the crosspiece so that the line measures 31 to 35 inches from one end to the other. When done, the crosspiece should easily slide up and down the spindle, and the line should fit within the spindle's notch.
- *Flywheel.* The flywheel sits on the lower end of the spindle, providing balance and weight. The ideal flywheel weight for creating an ember is 2½ to 3 pounds. To make one, gather two pieces of hardwood that are about 8 inches long by 2 inches wide, and burn or drill a hole into the exact center of each piece of wood. The size of the hole depends on the spindle's diameter. A proper fit allows the lower piece to wedge 4 inches up from the bottom of the spindle and the upper piece 2½ inches above that. Next, find two rocks that weigh about 1¼ to 1½ pounds each, and tie them between the two pieces of wood, one on each side of the hole, making sure that the holes are exposed and are in line with each other when done. Since it is doubtful that the rocks will weigh exactly the same, you may need to add twigs or line to one side of the flywheel to balance it out.
- *Fireboard.* The fireboard should be made from a material similar in hardness to the spindle plug. The ideal fireboard is 15 to 18 inches long, ¾ inch thick, and 2 to 3 inches wide. Prepare the fireboard by

Fire

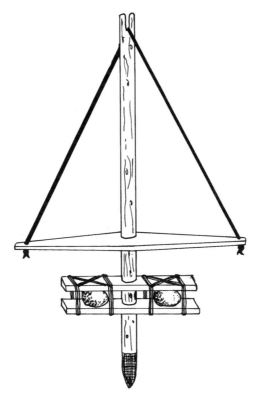

The various parts of a pump drill

carving a circular socket three-quarters the diameter of the spindle, at least 4 inches from one end, close to the long side (but not right on the side), and about one-quarter the thickness of the board.

Pump-drill technique
Place the plug inside the spindle and the flywheel over the top and into position. Advance the crosspiece down the spindle, letting the center of its line come to rest inside the upper notch. Place the spindle into the socket, and turn it by hand until the line is wrapped around it and the crosspiece has moved up the shaft. At this point, you are ready to begin. This system

is not much different from the bow and drill or hand drill. The spindle turns one way and then another while inside the fireboard's socket. The constant rotation creates friction, which in turn creates heat and eventually an ember. As with both of the other systems, your technique of execution will play a major role in success or failure.

Place one hand on each end of the crosspiece, kneel on your right knee, and place your left foot on the fireboard. Apply a smooth yet forceful downward stroke that has a rapid acceleration early on. As the crosspiece gets close to the bottom, don't relax. Continue to apply downward pressure while the line is rewrapping itself around the spindle in the opposite direction. By doing this, you maintain a friction force (contact) between the spindle and the fireboard. Try to time it so that as the crosspiece reaches the top, you can quickly accelerate back downward. Stop when a smooth, round indentation is made in the fireboard, and cut your notch in the same manner as described for the bow and drill. Once the notch is prepared, put bark, leather, or other dry material under your fireboard to catch the coal. Place the spindle back in the notch, and repeat the steps as just outlined until you have a solid wall of smoke and an ember is present.

Using the pump drill

Fine tuning the pump drill
- *Revolutions.* The ideal number of revolutions per complete cycle, top to bottom to top, is four—two down and two up. If you have more than that, the spindle loses its momentum when going up and, in turn, friction heat. Adjusting your string length will correct this problem.
- *Flywheel weight.* Although 2½ to 3 pounds is considered the ideal weight, this is not always true. To determine the best weight, use the following guidelines: If your spindle plug produces a dark brown dust and ember, it is just right. If it produces a light brown dust or no dust at all, it is too light. If the spindle is destroyed during use or it goes through the fireboard, the flywheel is too heavy.
- *Torque.* In order to obtain a fluid motion during operation, the torque applied during the downward motion of the crosspiece needs to approximate the torque created during its upswing. Adjusting the string's position on the crosspiece—equally on both sides—will correct this problem. You also may need to readjust the string's length to get the desired number of revolutions.

Fire plow

The fire plow is a method of rubbing two sticks together until an ember is created. This method is very difficult to master and takes a lot of practice. The best part about the fire plow is that no tools are required for its use and no notches need be created. Cottonwood and sotol are probably the two woods most often used for this technique. To make a fire plow, you need two pieces of wood: a plow and a fireboard.
- *Plow.* The plow—the piece you'll hold in your hand—should be about 1 foot long with one end ¼ to ½ inch wide.
- *Fireboard.* The fireboard needs to be about 2 inches wide and long enough for you to hold it in place while moving the plow across it.

Fire plow technique

Place the plow at a 90-degree angle to the base, and slowly begin pushing it back and forth, creating a groove that is 6 to 8 inches long. Place one hand close to the tip and the other at the butt end of the plow. Once you have a good groove established, lower the upper end of the plow, decreasing

The fire plow

its angle and creating an increased area of contact between the plow and fireboard. Once the set is smoking well, slightly raise the butt end of the plow back up. By doing this, you will intensify the heat on the tip. As you move the plow back and forth, make contact with the dust at least every other stroke. At first this may be difficult to do, and you may find yourself pushing the dust out of the groove completely. Practice. Sooner or later, you'll find an ember at the end of your groove.

Lighting tinder with an ember created from friction
Bark and grass are the most common tinders used with a friction heat source such as the bow and drill, hand drill, pump drill, or fire plow. Form a bird's nest with the tinder, and put it in a dry place where it is protected from the elements. Once you have created an ember with any of the friction methods, gently move it into the center of the bird's nest, and loosely fold the outer nest around it. Holding it all above your head, lightly blow on the ember, increasing in intensity until the tinder ignites. To avoid burning your fingers, it may be necessary to hold the tinder between two sticks. Once the tinder ignites, place it on the platform next to the brace, and begin building a fire.

Fire 75

Lighting tinder from a friction-based heat source

OXYGEN

Oxygen is necessary for the fuel to burn, and it needs to be present at all stages of a fire. To ensure this, you'll need a platform and brace. A platform can be any dry material that protects your fuel from the ground, such as dry tree bark or a dry, nonporous rock. Waterlogged rocks may explode

A platform and brace keep tinder dry and help ensure adequate oxygen flow to your fuel.

when heated; don't use them. A brace is usually a wrist-diameter branch that allows oxygen to circulate through the fuel when the fuel is leaned against it.

FUEL

Fuel can be broken down into three categories: tinder, kindling, and fuel. Each builds upon the previous one.

Tinder

Tinder is any material that will light from a spark. It's extremely valuable in getting the larger stages of fuel lit. Tinder can be man-made or natural.

Man-made tinder

When venturing into the wilderness, always carry man-made tinder in your survival kit. If you should become stranded during harsh weather conditions or a cold desert night, it may prove to be the key in having or not having a fire that first night. Since it is a one-time-use item, immediately start gathering natural tinder so that it can be dried out and prepared for use once your man-made tinder is used up. For natural tinder to work, it needs to be dry, have edges, and allow oxygen to circulate within it. For man-made tinder, this is not always the case; it may just need to be scraped or fluffed so it can catch a spark. The most common man-made forms of tinder are petroleum-based, compressed tinder tabs, and solid compressed fuel tablets.

Petroleum-based tinder

Petroleum-based tinder is very effective, even under harsh wet and windy conditions. Many kinds are available, but perhaps the most common is the Tinder-Quik tab. It is waterproof, odorless, and made from a light, compressible fiber impregnated with beeswax, petroleum, and silicones. To use it, simply fluff up the fiber so that it has edges to catch a spark. The tinder will burn for about two minutes. Although Tinder-Quik was designed for use with the Spark-Lite flint system, described above, it can be used with any heat source. Less expensive petroleum-based tinder can be made by saturating 100 percent cotton balls with petroleum jelly and stuffing them into a 35-millimeter film canister.

Fire

Compressed tinder tabs

WetFire tinder tablets are perhaps the most common compressed tinder tablets. They are waterproof, nontoxic, odorless, smokeless, and burn at about 1300 F for two to three minutes. Unlike the Tinder-Quik tabs, they are not compressible. To use, prepare the tinder by making a few small shavings to catch the sparks from your metal match.

Solid compressed fuel tablets

Besides serving as a heat source, these tablets easily light from a spark and can also be used as tinder to start a fire. More details can be found under Man-made Heat Sources, above.

Natural tinder

For natural tinder to work, it generally needs to be dry, have edges, and allow oxygen to circulate within it. Gather natural materials for tinder before you need it so that you have time to dry it in the sun, between your clothing, or by a fire. Remove any wet bark or pith before breaking the tinder down, and keep it off the damp ground during and after preparation. Some tinder will collect moisture from the air, so prepare it last and keep it dry until you're ready to use it. Natural tinder falls into three basic categories: bark; scrapings; and grass, ferns, leaves, and lichen. If you are uncertain if something will work for tinder, try it.

Bark

Prepare layered forms of tinder by working them between your hands and fingers until they're light and airy. To do this, start by holding a long section of the bark with both hands, thumb to thumb. Use a back-and-forth twisting action, working the bark until it becomes fibrous. Next, place the fibrous bark between the palms of your hands, and roll your hands back and forth until the bark becomes thin, light, and airy. At this point, you should be able to light it from a spark. Prepare this tinder until you have enough to form a small bird's nest. Place any loose dust created during this process in the center of the nest. Many types of bark will work as tinder, but birch is best as it will light even when wet due to a highly flammable resin it contains.

78   Surviving the Desert

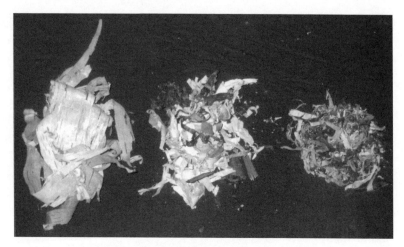

Bark being broken down

Making wood scrapings

Wood scrapings
Wood scrapings are created by repeatedly running your knife blade, at a 90-degree angle, across a flat section of pitch wood or heartwood. To be effective, you'll need enough scrapings to fill the palm of your hand. Like birch bark, pitch wood will light even when wet. The high concentration of pitch in the wood's fibers makes it highly flammable.

Grass, ferns, leaves, and lichen
As with bark, fashion a bird's nest with these materials. You may need to break them down further, depending on the materials at hand. This form of tinder needs to be completely dry to work successfully.

Leaves from an acacia tree make a great tinder.

80 Surviving the Desert

Kindling is usually composed of twigs or wood shavings.

Kindling

Kindling is usually composed of twigs or wood shavings that range in diameter from pencil lead to pencil thickness. It should easily light when placed on a small flame. Sources include small dead twigs found on the dead branches at the bottom of many trees; shavings from larger pieces of dry dead wood; small bundles of grass; and small pieces of dry sagebrush or cactus. You may also use heavy cardboard or gasoline- or oil-soaked wood.

Fuel

Fuel is any material that is thumb-size or bigger that will burn slowly and steady once lit. Kinds of fuel include dead tree branches or heartwood (the dry inside portion of a fallen tree trunk or large branch); dry dead cactus; dry dead sagebrush; dry dead grasses twisted into bunches; dry animal dung; and green wood that is finely split.

- *Dead cacti.* Once cacti are devoid of moisture, they provide an excellent fuel source and can quickly be broken down into the various stages of fuel. Cacti may be standing or lying on the desert floor.
- *Sagebrush.* Often live sagebrush has dead lower portions similar to the dead lower branches found on many trees. Breaking these branches away and then preparing them into the various stages of fuel is an easy process.
- *Dry grasses.* Dry grass is not only great tinder, but also provides an excellent fuel when tied into bundles. If this is your only source of fuel, tie the grass into bundles that are 12 to 24 inches long with varying diameters. This allows you to stage your fire from small to big.
- *Animal dung.* Because herbivores eat grass and other plants, their dried dung makes excellent fuel. Break the dung into various sizes to create tinder, kindling, and fuel.
- *Dry, dead branches at the bottom of trees.* This material is great during dry or very cold weather. It provides all the various stages of fuel when broken down properly. If the branches are wet, you'll

Dead sagebrush and shrubs provide an excellent fuel for building fires.

need to prepare it by scraping off all the wet bark and lichen. Run your knife across the wood's surface at a 90-degree angle. If it's still too wet, split the wood to expose its inner dry material.
- *Heartwood.* Heartwood requires a lot more energy and time when used to build a fire. However, it is ideal during wet conditions when you need a dry surface that will easily ignite. The best source is a stump with a sharp pointed top—in other words, a stump that wasn't created with a chain saw. Although probably not an issue in most desert climates, stumps that have a flat surface can absorb massive amounts of moisture, especially when capped with snow. To gather heartwood, pull, kick, or rip pieces off the stump. If you're unable to separate the wood from the stump, wedge a sturdy pole between the main stump and a loose piece of wood, or use your large fixed-blade knife to help it along. Once gathered, break the pieces of wood down from large to small.
- *Green wood.* If you have a hot fire, green wood that is finely split will burn. However, it should not be used in the early stages of your fire. To increase your odds of success, remove the outer bark and cambium layer.

STEPS TO BUILDING A FIRE
When building a fire, it's important to gather enough fuel to build three knee-high fires. This allows you to go back to a previous stage if the fire starts to die and to keep the fire going while you get more material. Once the wood or other fuel is gathered, break it down from big to small, always preparing the smallest stages last. This will help decrease the amount of moisture your tinder and kindling collect during the preparation process. If conditions are wet, you'll need to strip off all lichen and bark, and for best results, split the branches in half to expose the dry inner wood. Construct a platform and brace (described under Oxygen above), and use the brace to keep your various stages of fuel off the ground while breaking it down.

Once all the stages of fuel are prepared, either light it or place the lit tinder on the platform next to the brace. Use the brace to place your smaller kindling directly over the flame. Spread a handful of kindling over the flame all at once, instead of one stick at a time. Once the flames lick up through the kindling, place another handful perpendicularly across the

The various stages of fire using the desert acacia tree

first. When this stage is burning well, advance to the next size. Continue crisscrossing your fuel until the largest size is burning and the fire is self-supporting. If you have leftover material, set it aside in a dry place so that it can be used to start another fire later. If you have a problem building your fire, reevaluate your heat, oxygen, and fuel to determine which one is not present or is inadequate for success.

FIRE REFLECTOR
Consider building a firewall to reflect the fire's heat in the direction you want. Secure two 3-foot-long poles into the ground 1 foot behind the fire circle. In order to pound the poles into the ground, you'll need to sharpen the ends and use a rock or another sturdy pole to safely drive them in. Next, place two more poles of similar size 4 to 6 inches in front of the first ones. Gather green logs of wrist diameter, and place them between the

84   Surviving the Desert

Fire reflector

poles to form a 3-foot-high wall. You can lean the wall slightly to reflect the heat down or up.

MAINTAINING A HEAT SOURCE
Several methods are commonly used to maintain a heat source for ongoing or later use.

KEEPING A FLAME
The best way to keep a flame is to provide an ongoing fuel source. The type of fuel you use is important. Softwoods, such as cedar, pine, or fir, provide an excellent light and heat source, but they burn up rather fast. Hardwoods, such as maple, ash, oak, or hickory, will burn longer and produce less smoke. These woods are ideal for use at night. In the desert, however, your fuel may be limited to dead cactus, sagebrush, dry grass, and animal dung. All of these fuels burn hot and fast.

KEEPING A COAL
Either banking the fire or storing it inside a fire bundle can maintain a coal.

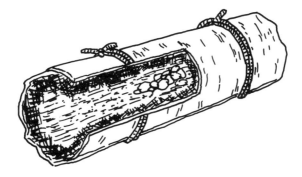

Fire bundle

Banking the fire

If you are staying in one place, bank the fire to preserve its embers for use at a later time. Once you have a good bed of coals, cover them with ashes and/or dry dirt. If done properly, the fire's embers will still be smoldering in the morning. To rekindle the fire, remove the dry dirt, lay tinder on the coals, and gently fan the coals until the tinder ignites.

Fire bundle

If you plan on traveling, you can use a fire bundle to transport the coal. A fire bundle can keep a coal for six to twelve hours. To construct it, surround the live coal with dry punk wood or fibrous bark such as cedar or juniper, and wrap this with damp grass, leaves, or humus. Around this, wrap a heavy bark such as birch. The key to success is to ensure that there is enough oxygen to keep the ember burning, but not enough to promote its ignition. If the bundle begins to burn through, it may be necessary to stop and build another fire from which to create another coal for transport.

SURVIVAL TIPS

BE PREPARED FOR COLD NIGHTS

Temperature extremes are common in deserts. Although the days may be hot, nights can become quite cold. Plan to build a fire to help take the chill off that evening air. Once you've finished building a shelter, take the time to collect fire-building materials. You'll be happy you did.

7

Signaling

During an average year, the U.S. National Park System conducts around four thousand search-and-rescue operations. Of these, approximately half of the missions involve a seriously injured or ill subject, and in 5 percent of the cases the victim dies.

RULES OF SIGNALING
A properly used signal increases a survivor's chances of being rescued. A signal has two purposes: to attract rescuers to your whereabouts, and then to help them pinpoint your exact location. When preparing a signal, use the following rules:

- *Stay put.* Once you realize you're lost, stay put. Depart only if the area you are in doesn't meet your survival needs, rescue is not imminent, and you know where you are and have the navigational skills to get to where you want to go. If you are lost or stranded in a car, plane, or ATV, stay with it; the vehicle will serve as a ground-to-air signal. When a search is activated, rescuers will begin looking for you in your last known location. If for some reason you need to move, leave a ground-to-air signal pointing in your direction of travel, along with a note telling rescuers of your plans. If you do move, go to high ground and find a large clearing from which to signal.
- *Properly locate your signal site.* Your signal site should be close to your camp or shelter, located in a large clearing that has 360-degree visibility, and free of shadows.
- *Use one-time-use signals at the appropriate time.* Many signals are one-time-use items and thus should be ignited only when you see or hear a potential rescuer and are sure he or she is headed in your direction.

- *Know and prepare your signal in advance.* Since seconds can mean the difference between life and death, don't delay in preparing or establishing a signal.

SIGNALS THAT ATTRACT RESCUE

The most effective distress signals for attracting attention are aerial flares and parachute flares, because they are moving, spectacular, and cover a large sighting area. When using an aerial or parachute flare, you need to adjust for any drift created by the wind. Since you want the flare to ignite directly overhead, you'll need to point the flare slightly into the wind, usually about 5 to 10 degrees.

AERIAL FLARE

An aerial flare is a one-time-use item and should be used only if a rescue team, aircraft, or vessel is sighted. As with all pyrotechnic devices, it is flammable and should be handled with caution. Most aerial flares fire by pulling a chain. In general, you'll hold the launcher so that the firing end—where the flare comes out—is pointed overhead and skyward, allowing the chain to drop straight down. Next, while the flare is pointed skyward, use your free hand to grasp and pull the chain sharply downward. Make sure the hand holding the launcher is located within the safe area as detailed on the device you are using. For safe use and best results, hold the flare away from your body and perpendicular to the ground. The average aerial flares have a 500-foot launch altitude, six-second burn time, and 12,000 candlepower. Under optimal conditions, these flares have been sighted up to 30 miles away. Many aerial flares float and are waterproof. Their average size is 1 inch in diameter and about $4\frac{1}{2}$ inches long when collapsed. Depending on your needs, you can purchase disposable flares or ones that allow replacement cartridges. The Orion Star-Tracer and the SkyBlazer XLT aerial flares are two good examples and can be found in most sporting-goods or marine stores.

PARACHUTE FLARE

A parachute flare is simply an aerial flare attached to a parachute. The parachute allows for a longer burn time while the flare floats down to earth. A parachute flare is a one-time-use item and should be used only if a rescue team, aircraft or vessel is sighted. It is flammable and should be handled with caution. The Pains Wessex SOLAS Mark 3 parachute flare can reach a height of 1,000 feet and produce a brilliant 30,000 candlepower. The flare's red light drifts down to earth under a parachute and has a burn time of about forty seconds.

SIGNALS THAT PINPOINT YOUR LOCATION

Once help is on the way, handheld red signal flares, orange smoke signals, signal mirrors, kites, strobe lights, whistles, and ground-to-air signals can serve as beacons to help rescuers pinpoint your position and keep them on course.

HANDHELD RED SIGNAL FLARE

A handheld red signal flare is a one-time-use item and should be used only if a rescue team, aircraft, or vessel is sighted. It is flammable and should be handled with caution. To use one, stand with your back to the wind, and keep the device pointed away from your face and body during and after lighting. Most red signal flares are ignited by removing the cap and striking the ignition button with the cap's abrasive side. To avoid burns, hold the flare in its safe area, and never hold it overhead. Most devices will burn for two minutes, have a 500 candlepower, and are about 1 inch in diameter by 9 inches long. For increased burn time and candlepower, you might consider getting a handheld marine red signal flare, which averages a burn time of three minutes and a 700 candlepower. These devices work best when used at night.

ORANGE SMOKE SIGNALS

An orange smoke signal is a one-time-use item and should be used only if a rescue team, aircraft, or vessel is sighted. It is flammable and should be handled with caution. To use one, stand with your back to the wind, and keep the device pointed away from your face and body during and after lighting. Other than wind, snow, or rain, the biggest problem associated with a smoke signal is that cold air keeps the smoke close to the ground, sometimes dissipating it before it reaches the heights needed to be seen. Two types are the SkyBlazer and the Orion.

SkyBlazer orange smoke signal

The SkyBlazer smoke signal is about the size of a 35-millimeter film container and thus is easy to carry. It's also easy to use, and the directions are on the container. Simply remove its seal, pull the chain, and then place it on the ground—the SkyBlazer smoke signal is not a handheld device. The signal lasts for only forty-five seconds under optimal conditions and produces only a small amount of orange smoke. In order to create a more appropriate quantity of smoke, I have used two at once.

Orion handheld orange smoke signal

The Orion signal is bigger than the SkyBlazer. It comes in two sizes: marine and wilderness. The marine signal is about the size of a road flare, and the wilderness signal is half that. The Orion, too, has easy-to-read directions right on the signal. Simply remove the cap, and then strike the ignition button with the abrasive part of the cap. To avoid burns, hold the flare in its safe area, and never hold it overhead. These signals put out a lot of smoke and last more than sixty seconds. If space permits, this is a far more effective signal than the SkyBlazer. Orion also makes a floating orange smoke signal that lasts for four minutes.

SIGNAL MIRROR (WITH SIGHTING HOLE)

On clear, sunny days, signal mirrors have been seen from as far away as 70 to 100 miles. Although the signal mirror is a great signaling device, it requires practice to become proficient in its use. Most signal mirrors have directions on the back, but here are general guidelines on how to use one. Holding the signal mirror between the index finger and thumb of one hand, reflect the sunlight from the mirror onto your other hand. While maintaining the sun's reflection on your free hand, bring the mirror up to eye level and look through the sighting hole. If done properly, you should see a bright white or orange spot of light in the sighting hole. This is commonly called the aim indicator or fireball. Holding the mirror close to your eye, slowly turn it until the aim indicator is on your intended target. If you lose sight of the aim indicator, start over. Since the mirror can be seen from a great distance, sweep the horizon throughout the day, even if no rescuers are in sight. On land, add movement to the signal by slightly wiggling the mirror.

90 Surviving the Desert

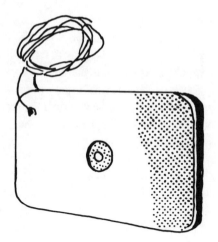

Signal mirror

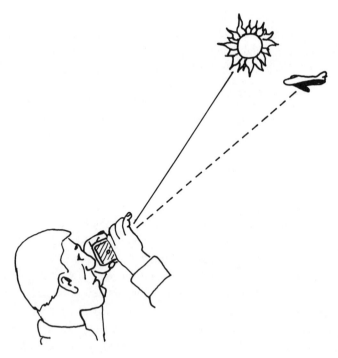

Using a signal mirror

Signaling 91

The flash from a signal mirror as seen by a rescue helicopter

At sea, hold the mirror steady so that it stands out from the sparkles created by water movement. If signaling an aircraft, stop after you're certain the pilot has spotted you, as the flash may impede his or her vision.

KITE

A kite is a highly visible signal that not only attracts attention to your location, but also helps rescuers pinpoint where you are. David Instruments' Sky-Alert Parafoil Rescue Kite is a good example. The 28-by-38-inch kite flies in 5 to 25 knots of wind and requires only about 8 to 10 knots to lift another signaling device such as a strobe or handheld flare. A benefit of this signal is that it can be working for you while you attend to other needs. In addition to providing a great signal, flying the kite can also help alleviate stress.

STROBE LIGHT

A strobe light is a device that fits in the palm of your hand and provides an ongoing intermittent flash. ACR Electronics' Personal Rescue Strobe is a good example. It delivers a bright flash (250,000 peak lumens) at one-second intervals and can run up to eight hours on AA batteries. It is visible for up to one nautical mile on a clear night. As with all battery-operated devices, strobe lights are vulnerable to cold, moisture, sand, and heat; protect the strobe from these hazards by any means available.

WHISTLE

Always carry a whistle on your person. A whistle will never wear out, and its sound travels farther than the screams of the most desperate survivor. If you become lost or separated, immediately begin blowing your whistle in multiple short bursts. Repeat every three to five minutes. If rescue doesn't appear imminent, go about meeting your other survival needs, stopping periodically throughout the day to blow the whistle. It may alert rescuers to your location, even if you're unaware of their presence. Storm Whistle's Storm Safety Whistle is a good example. Its unique design makes it the loudest whistle you can buy, even when soaking wet. It's made from plastic and has easy-to-grip ridges.

GROUND-TO-AIR PATTERN SIGNAL

A ground-to-air signal is an extremely effective device that allows you to attend to your other needs while continuing to alert potential rescuers to your location. Although you can buy a signal panel, I'd suggest purchasing a 3-by-18-foot piece of lightweight nylon—orange and white. There are three basic signal designs you should know, and each can be made using the nylon:

    V = Need assistance
    X = Need medical assistance
    ↑ = Proceed this way

Once you've created the appropriate signal, stake it out so that it holds its form and doesn't blow away. For optimal effect, follow these guidelines:

- *Size.* The ideal signal has a ratio of 6:1, with its overall size at least 18 feet long by 3 feet wide.

- *Contrast.* The signal should contrast the surrounding ground cover, such as orange on white ground and white on brown or green.
- *Angularity.* Because nature has no perfect lines, a signal with sharp angles will be more effective.
- *Shadow.* In summer, elevate the signal. In winter, stomp or dig an area around the signal about 3 feet wide. If the sun is shining, either of these methods will create a shadow, which ultimately increases the signal's size.
- *Movement.* Setting up a flag next to your signal may create enough movement to catch the attention of a rescue party. It is also advisable

Ground-to-air signal as seen by a rescue helicopter

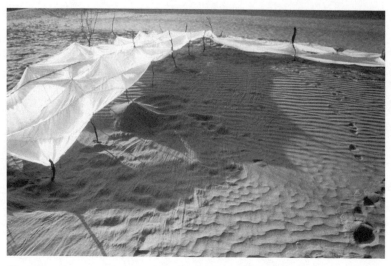

Elevating the ground to air signal creates a shadow and makes it appear larger.

to suspend a flag high above your shelter so that it can be seen from all directions by potential rescuers.

CELLULAR PHONES

Although a cellular phone is a great thing to have, it's not without limitations and often doesn't work in remote areas. Do not rely on one as your sole signaling and rescue device. Not only are cell phones limited by their service area, but they are also vulnerable to sand, heat, cold, and moisture. You'll need to protect the phone from these hazards by any means available.

IMPROVISED SIGNALS

Many manufactured signals are one-time-use items or are limited by their battery life, and it may be necessary to augment them with an improvised signal. A fire can be as effective as a red flare; a smoke generator works better and lasts longer than an orange smoke signal; an improvised signal mirror can be as useful as a manufactured one; and a ground-to-air signal can be made from materials provided by Mother Nature.

FIRE AS A SIGNAL

During the night, fire is probably the most effective means of signaling available. One large fire will suffice to alert rescuers to your location. Don't waste time, energy, and wood building three fires in a distress triangle unless rescue is uncertain. Prepare the wood or other fuel for ignition prior to use.

SMOKE GENERATOR

Smoke is an effective signal if used on a clear, calm day. If the weather is bad, however, chances are the smoke will dissipate too quickly to be seen. The rules for a smoke signal are the same as those for a fire signal: You only need one, and prepare the materials for the signal in advance. To make the smoke contrast against its surroundings, add any of the following materials to your fire:

- *To contrast against sand or other light backgrounds:* Use tires, oil, or gasoline to create black smoke.
- *To contrast against darker backgrounds:* Use boughs, grass, green leaves, moss, ferns, or even a small amount of water to create white smoke.

Set up the smoke generator in advance so that it can be quickly lit when rescuers are spotted. To do this, build, but don't light, a tepee firelay with a platform and brace, using a lot of tinder and kindling in the process. Then construct a log-cabin-style firelay around the tepee, using any fuel that is thumb-size and larger. Leave a small, quick-access opening that will allow you to reach the tinder when it comes time to light it. Finally, cover the whole thing with heavy boughs or similar material depending on the desert you are in. This covering should be thick enough to protect the underlying structure from the elements. When done, the generator should look like a haystack. Once rescuers have been spotted or appear to be headed in your direction, gently pick up one side of the covering and light the smoke generator. If you have trouble getting it lit, this is one of the rare circumstances when I'd advise using your red smoke flare as a heat source.

IMPROVISED SIGNAL MIRROR

You can create a signal mirror from anything shiny, such as a metal container, coin, credit card, watch, jewelry, or belt buckle. Although an improvised signal mirror can make a great signaling device, it requires practice to become proficient in its use. To use one, follow these steps: Holding the device between the index finger and thumb of one hand, reflect the sunlight from the mirror onto the palm of your other hand. While keeping the reflection on that hand, create a V between your thumb and index finger. Move the light reflection and your hand until the rescue aircraft or other rescuer is in the V. At this point, move the reflected light into the V and onto your intended target. Since the mirror can be seen from great distances, sweep the horizon periodically throughout the day, even if no rescue vehicles are

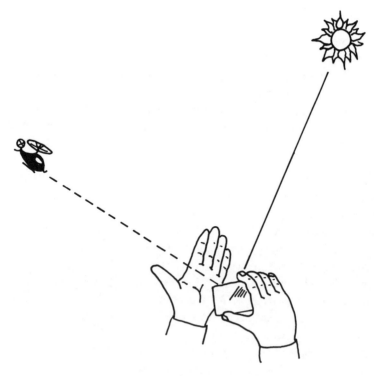

Improvised signal mirror

in sight. On land, wiggle the mirror slightly to add movement to the signal. At sea, hold the mirror steady to contrast with the sparkles created by the natural movement of the water. When signaling an aircraft, stop flashing after you're certain the pilot has spotted you, as it may impede his or her vision.

IMPROVISED GROUND-TO-AIR PATTERN SIGNAL
If you don't have a signal panel, you can improvise one from what Mother Nature provides—boughs, bark, logs, cacti, grass, sagebrush, or any other material that contrasts the ground color. Follow the basic guidelines of construction under Ground-to-Air Pattern Signal above.

HELICOPTER RESCUE
Helicopter rescues are becoming more frequent as more and more people head into the wilderness. Rescue crews may be civilians, but more often than not they are either military personnel or Coast Guard. If the helicopter can land, it will. If not, a member of the rescue team will be lowered to your position. At this point, you'll either be hoisted to the helicopter or moved to a better location while in a harness or basket and dangling from the helicopter. Secure all loose items before the helicopter lands, or they may be blown away or sucked up into the rotors. Once the helicopter has landed, do not approach it until signaled to do so, and only approach from the downhill front side. This will ensure that the pilot can see you and decrease your chances of being injured or killed by the rotor blades.

SURVIVAL TIPS

SIGNAL OFTEN
Since you never know if someone might be in the vicinity, blow your whistle every five to ten minutes, and scan the horizon periodically with your signal mirror.

ALWAYS BE PREPARED
To avoid watching rescuers disappear while you're still fumbling with your signals, learn how to use them in advance, and make sure they are ready to use before rescue is near.

8
Water

In the desert, the lack of water plays a major role in survivability. It is an extremely valuable resource and is far more important than food. You can live anywhere from three weeks to two months without food, but only days without water. Thus you need to pack in enough water for at least one day, along with the necessary equipment to procure more. Check with the local authorities on accessible water related to your trip before departing. Don't rely on someone who hasn't been in the area recently or is giving you hearsay information. It's better to adjust or delay your trip than to run out of water.

Our bodies are composed of approximately 60 percent water. Looking at this in more detail, our brains are composed of about 70 percent water, our blood 82 percent, and our lungs 90 percent. In the bloodstream, water helps metabolize and transport vital elements, carbohydrates, and proteins necessary to fuel our bodies. Water also helps us dispose of our bodily waste. In order for our bodies to continue normal operations, water lost through perspiration, respiration, digestion, urination, and defecation needs to be replaced. Water plays a vital role in our ability to get through a day, and it's hard to understand why so many people drink so little.

During a normal, nonstrenuous day, your body needs 2 to 3 quarts of water. When physically active or in extreme hot or cold environments, however, that need increases to at least 4 to 6 quarts a day. It's important to drink enough water to prevent dehydration. A person who's mildly dehydrated develops excessive thirst and becomes irritable, weak, and nauseated. As the dehydration worsens, the person shows a decrease in mental capacity and coordination, and it becomes difficult to accomplish even the simplest of tasks. The importance of water in a hot environment cannot be overemphasized, and its availability—carried on you or otherwise—must be continually assessed during any trip into the desert.

The best storage container for your water is your body. If water becomes scarce, do not ration it. Instead, follow these basic guidelines to limit water loss from sweat:
- Work in the evening and morning, when temperatures are lower.
- During the day, rest in a shaded area.
- Wear protective clothing in a loose and layered fashion, with a wide-brimmed hat and neck covering (described in detail in chapter 4).

DISPELLING MYTHS ABOUT WATER

NEVER DRINK URINE!
By the time you even think about doing this, you are very dehydrated. That means your urine is full of salts and other waste products. For a hydrated person, urine is 95 percent water and the rest is waste products such as urea, uric acid, and salts. As you become dehydrated, the amount of water decreases, and the concentration of salts increases substantially. If you drink these salts, the body will draw upon its water reserves to help eliminate them, and thus you will actually lose more water than you might gain from your urine.

NEVER DRINK SALT WATER!
Salt water often has salt concentrations even higher than those found in urine. Here again, if you drink these salts, the body will draw upon its water reserves to help eliminate them, and thus you will actually lose more water than you gain.

NEVER DRINK BLOOD!
Blood is composed of plasma, red and white blood cells, and platelets. Plasma, about 55 percent of the blood's volume, is predominantly water and salts and also carries a large number of important proteins and small molecules, such as vitamins, minerals, nutrients, and waste products. Waste products produced during metabolism, such as urea and uric acid, are carried by the blood to the kidneys, where they are transferred from the blood

into urine and eliminated from the body. The kidneys carefully maintain the salt concentration in plasma. If you drink blood, you are basically ingesting salts and proteins, and the body will draw upon its water reserves to help eliminate them. Again, you will lose more water than you might gain.

WATER INDICATORS

In hot environments, water may be difficult to find. Map indicators may be misleading, since most desert water is intermittent. Map markings that represent intermittent streams and springs are often relevant only after it has rained, and the windmills, tanks, and troughs shown may be broken down or dry. Understanding water indicators created by plants and trees, insects, birds, mammals, and the terrain will be helpful when trying to find water.

PLANTS AND TREES

Plush green vegetation found at the base of a cliff or mountain may indicate a natural spring or underground source of water. Most deciduous trees require large amounts of water to survive and often indicate the presence of surface water or groundwater.

INSECTS

If bees are present, water is usually within several miles of your location. Ants require water and often place their nest close to a source. Swarms of mosquitoes and flies are a good predictor that water is close.

BIRDS

Birds frequently fly toward water at dawn and dusk in a direct, low flight path. This is especially true of birds that feed on grain, such as pigeons and finches. Flesh-eating birds may also exhibit this flight pattern, but their need for water isn't as great, and they don't require as many trips to the water source. Birds observed circling high in the air during the day are often doing it over water as well.

MAMMALS

Like birds, mammals frequently visit watering holes at dawn and dusk. This is especially true of mammals that eat a grain or grassy type of diet.

Water   101

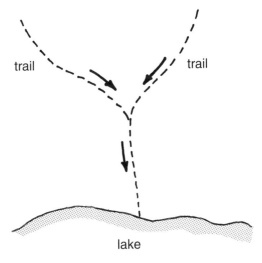

Two trails merging together often point toward water.

Watching their travel patterns or evaluating mammal trails may help you find a water source. Trails that merge into one are usually a good pointer, and following the merged trail often leads to water.

TERRAIN
Drainages, valleys, and ground depressions are good water indicators and may hold surface water or groundwater, especially after a rain. Often springs can be found at the base of a plunge pool (dry waterfall) or the bend of a dry riverbed with lush green vegetation.

NATURAL WATER SOURCES
Since your body needs a constant supply of water, you'll eventually need to procure water from Mother Nature. Various sources include surface water, groundwater, precipitation, and plants.

SURFACE WATER
Surface water may be obtained from rivers, creeks, springs, basin lakes, and rock depressions. These sources may be found year-round, but in desert

102   Surviving the Desert

environments they typically are present only when there has been a recent rainfall. Most sources are stagnant or slow-moving and prone to contamination from protozoans, bacteria, and viruses. They should all be treated.

Rivers and creeks

Rivers often support deciduous trees and other vegetation not otherwise found in a desert. When you spot an area with vegetation, especially plants or trees that run in a line, you should investigate—surface water might be there, too. Creeks, on the other hand, are often dry, and finding water there may be a difficult task. Look for areas of lush green vegetation, which is

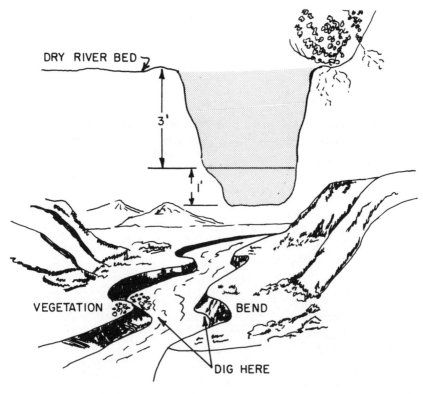

Dry riverbed

often found downstream and at the bend of a pool. Water may be on the surface here, or you might find groundwater by digging down 2 or 3 feet.

Springs
Springs occur when the water table crosses the ground's surface and are often dependent on rains. In other words, if it hasn't rained for some time, it is doubtful water will be present. It is hard to know where a spring will occur. However, they are often found in low-lying areas next to a hillside and support lush green vegetation.

Basin lakes
Basin lakes occur where water is trapped without any means of escape. They can appear in low-lying areas or when intermittent creeks or rivers flow into an area that provides no exit. Water may or may not be present, depending on when the last rainfall was.

Rock depressions
If there has been a recent rain, water is often trapped in rock depressions. This water is often stagnant, but it nevertheless is an option that should be considered in a desert climate. As with most water, it should be purified before consumed.

GROUNDWATER
Groundwater is found under the earth's surface. This water is naturally filtered as it moves through the ground and into underground reservoirs known as aquifers. Although treatment may not be necessary, always err on the side of caution. Locating groundwater is probably the most difficult part of accessing it. Look for things that seem out of place, such as a small area of lush green vegetation at the base of a hill or a bend in a dry riverbed that is surrounded by sparser green or even brown vegetation. A marshy area with a fair amount of cattail or hemlock growth may provide a clue that groundwater is available. I have found natural springs in desert areas and running water less than 6 feet below the earth's surface using these clues. For ease of access, you can dig a small seepage basin well at the source, following the directions under Natural Water Filtration below. If you are on a

coastal desert, you can procure fresh water by digging such a well one dune inland from the beach.

PRECIPITATION

The two forms of precipitation in a hot desert climate are occasional rain and dew. If it rains, set out containers or dig a small hole and line it with plastic or other nonporous material. After the rain has stopped, look for water in crevasses, fissures, and low-lying areas. If morning dew is present, use a porous cloth to absorb it, then wring out the moisture into your mouth.

PLANTS

Depending on the type of desert and time of year, you might be able to get some water from plants. Solar stills that collect moisture from plants in the form of condensation can provide some water in hot climates and should be considered. These include the vegetation bag and transpiration bag. Contrary to popular belief, most cacti (including the saguaro cacti) do not provide an unending water source when cut into. Often a moist pulp is present, and a small amount of water can be obtained by squeezing it inside a porous material.

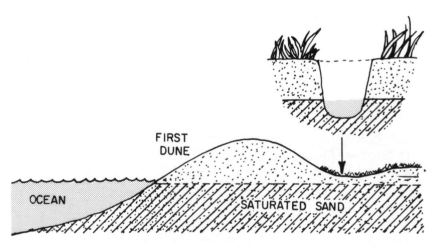

Beach well water source

Don't rely on the saguaro cactus to meet your water needs.

Vegetation bag

To construct a vegetation bag, you need a clear plastic bag and an ample supply of healthy, nontoxic vegetation. A 4- to 6-foot section of surgical tubing is also desirable. Open the plastic bag and fill it with air to make it easier to place the vegetation inside the bag. Fill the bag one-half to three-quarters full of lush green vegetation, taking care not to puncture it. Place a small rock or similar item into the bag, and if you have surgical tubing, slide one end inside and toward the bottom of the bag, tying an overhand knot in the other end. Close the bag and tie it off as close to the opening as possible. Place it on a sunny slope so that the opening is on the downhill side and slightly higher than the bag's lowest point. Position both the rock and surgical tubing at the lowest point in the bag. Drain off all liquid prior to sunset, or it will be reabsorbed into the vegetation. If using surgical tubing, simply untie the knot and drink the water. If no tubing is used,

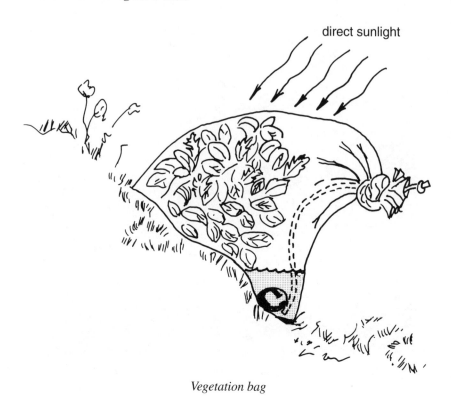

Vegetation bag

loosen the tie and drain off the available liquid. For best results, change the vegetation every two to three days.

Transpiration bag
The advantage of a transpiration bag over a vegetation bag is that the same vegetation can be reused, after allowing enough time for it to rejuvenate. To construct a transpiration bag, you need a clear plastic bag and an accessible, nonpoisonous tree or brush. A 4- to 6-foot section of surgical tubing is also desirable. Open the plastic bag and fill it with air, then place it over the leafy vegetation of the tree or brush on the side with the greatest exposure to the sun. Take care not to puncture the bag. Place a small rock or similar item into the bag's lowest point, and if you have surgical tubing, place one end at the bottom of the bag next to the rock, tying an overhand knot in the

other end. Close the bag and tie it off as close to the opening as possible. Drain off all liquid prior to sunset, or it will be reabsorbed into the plant. If using surgical tubing, simply untie the knot and drink the water. If no tubing is used, loosen the tie and drain off the available liquid. Change the bag's location every two to three days to allow the previous site to rejuvenate so it might be used again later.

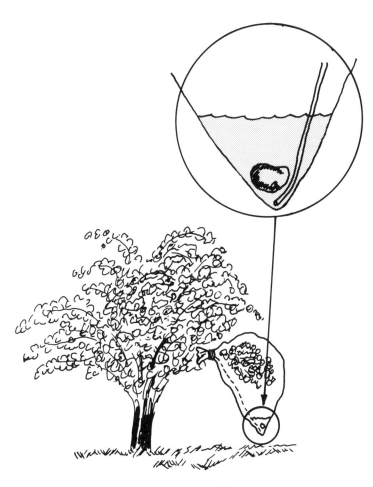

Transpiration bag

MAN-MADE WATER SOURCES

Since water is often a scarce commodity in desert climates, mankind has derived multiple procurement, transport, and storage systems where it is found. These include windmills, storage tanks, pipelines, spring boxes, and man-made dams. Water from any source should be purified and, when appropriate, filtered.

WINDMILLS

A water pump windmill, still common in many rural areas of the United States, uses the energy from wind to draw water from underground. These machines have several obliquely angled blades mounted on a descending horizontal shaft and a fantail rudder that steers the blades into the wind. When the windmill's blades turn, the energy is transmitted downward through a system of horizontal shafts and gears that operate a piston pump. These pumps use suction to transport water from underground to the earth's surface. Windmills are also used to help move water through pipes from the source to another location. When wind velocities become excessive, safety devices automatically turn the rotor out of the wind to prevent damage to the mechanism.

If you come across a windmill, take the time to check it out. Although many have been abandoned, it may still provide a water source. However, don't get your hopes up. Most active sites have signs that maintenance has occurred, and there may be grease on the moving parts, newer paint, or a storage area with modern tools. Is the windmill over the water source, or is it part of a transport system? If it appears to be the original pump, look for a storage tank. In either situation, look for a smaller water container that might be kept full in case the pump needs to be primed during maintenance. The windmill will not work if the brake is on, so check to see if it is on or off. When on, the windmill's tail will be folded next to the fan, keeping it out of the wind. When off, the tail will be extended and cause the fan to turn into the wind. The brake lever is normally located at the base of the windmill. If a wind is present and you release the brake, an operational windmill should begin to work. If it doesn't, look for a hand pump or consider turning the blades by hand. If you decide to climb the tower to manually turn the blades, survey it well beforehand. If you fall and get hurt, your situation will be worse. If the fan is rusted or rotates but

Water pump windmills are still common in many rural areas.

nothing happens, try hitting the windmill's shaft (not the casing) with a rock or similar object. Doing this might free it up from rust or dislodge a rock that is preventing it from working.

STORAGE TANKS AND PIPELINES
Storage tanks and pipelines can often be found next to windmills or other energy sources. These tanks and lines often have open and close valves, plugs, and spigots on the outlet pipe. If you can't access the water via a spigot, you may have trouble getting it from a valve or plug source without a wrench. If all else fails and your survival is in question, try to break the line by whatever means you can.

110   Surviving the Desert

Pipelines can often be found next to windmills and are a potential water source.

SPRING BOXES AND MAN-MADE DAMS

A spring box is a natural spring with a rock or concrete box built around it. Usually the source is covered, with pipelines carrying the water to a storage system. Man-made dams are often created in creek beds and drainages where rain runoff can be collected.

<u>WATER FILTRATION</u>

Filtration systems do not purify water. At best, they remove unwanted particles and make the water more palatable.

SEEPAGE BASIN

A seepage basin is one method of filtering water, although it does not always take away the water's awful taste. This method is not ideal for a desert climate, and the amount of sweat lost constructing the basin makes it a bad choice. If you still wish to create a seepage basin well, dig a 3-foot-wide hole about 10 feet from your water source. Dig down until water begins to seep in, and then go about another foot. Line the sides with wood or rocks so that no more dirt or sand will fall in, and let it sit overnight.

LAYERED FILTER

Another method of filtration is to use a layered filtering device, running water through grass, sand, and black charcoal. This does more than just remove unwanted particles; it also makes the water taste better. This system can be created using a three-tiered tripod design or by simply layering the materials inside a container such as a large coffee can. For a three-tiered tripod, tie three sections of porous material about 1 foot apart. Fill the top one with grass, the middle with sand, and the bottom with charcoal. To use, pour the water into the top filter, and catch it in a container as it comes out the bottom.

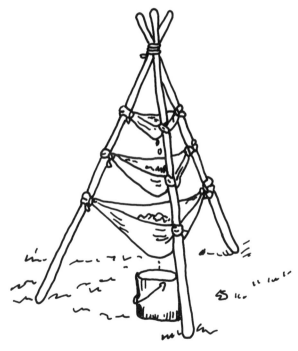

Three-tiered tripod

SIMPLE FILTER
Perhaps the quickest and easiest method to filter water in a desert climate is to simply pour the water through a porous material. Although this will not remove a bad taste, it will get rid of larger sediment.

WATER IMPURITIES
According to the Centers for Disease Control (CDC), water contaminated with microorganisms causes over a million illnesses and a thousand deaths in the United States each year. The primary disease-causing organisms, or pathogens, fall into three categories: protozoans (including cysts), bacteria, and viruses. Contaminants may also pose a problem.

PROTOZOANS
Protozoans are one-celled organisms that vary in size from 2 to 100 microns, live in many insects and animals, and survive in cysts (protective shells) when outside of an organism. They include *Giardia* and *Cryptosporidium*. It takes only a few organisms to infect someone, and once inside a host, protozoans rapidly reproduce, causing severe diarrhea, abdominal cramps, bloating, fatigue, and weight loss.

BACTERIA
Bacteria can be as tiny as 0.2 microns, much smaller than protozoans. They include typhoid, paratyphoid, dysentery, and cholera. Bacteria are often present in both wild and domestic animals. Once in the water, they can survive for weeks, even longer if frozen in ice.

VIRUSES
Viruses can be as small as 0.004 microns, which makes it easy for them to pass through a filter. Viruses found in water include hepatitis A and E, Norwalk virus, rotavirus, echovirus, and poliovirus. Unlike protozoans and bacteria, there is no treatment for a waterborne viral infection, and this makes them a significant health risk, especially for people with a compromised immune system.

CONTAMINANTS
Other impurities that can be found in water include disinfectants and their byproducts, inorganic and organic chemicals, and radionuclides.

PURIFYING WATER

There are three basic methods for treating your water: boiling, chemical treatments, and commercial filtration systems. Boiling is far superior to chemical treatments and should be done whenever possible.

Boiling

To kill any disease-causing microorganisms, the Environmental Protection Agency (EPA) recommends a vigorous boil for one minute. My rational mind tells me that this must be based on science and should work. After seeing one of my friends lose about 40 pounds from a severe case of giardiasis, however, I tend to boil it longer.

Chemical treatment

When unable to boil your water, you may elect to use chlorine or iodine. These chemicals are effective against bacteria, viruses, and *Giardia,* but according to the EPA, there is some question about its ability to protect you against *Cryptosporidium*. In fact, the EPA advises against using chemicals to purify surface water. If you must, chlorine is preferred over iodine, since it seems to offer better protection against *Giardia*. Combining a chemical treatment with a commercial filtering device will make your water safe to drink. Both chlorine and iodine tend to be less effective in cold water.

Household chlorine bleach

The amount of bleach to use for purifying water depends on the percentage of available chlorine in the solution. This can usually be found on the label.

Available Chlorine	Drops per Quart of Clear Water
1%	10
4-6 %	2
7-10 %	1
unknown	10

If the water is cloudy or colored, double the recommended amount of bleach. Put the water and bleach in a container with a lid, wait three minutes, and then vigorously shake the container with the lid slightly loose, allowing some water to seep out. Seal the lid and wait another twenty-five to thirty minutes, then loosen the lid and shake the container again. At this point, consider the water safe to consume provided there is no *Cryptosporidium* in the water.

Iodine
There are two forms of iodine that are commonly used to treat water: tincture and tablets. The tincture is nothing more than the common household iodine that you may have in your medical kit. This product is usually a 2 percent iodine solution, and you'll need to add five drops to each quart of water. For cloudy water, double this amount. Mix the iodine into the water and let it stand for thirty minutes. With iodine tablets, use one tablet per quart of warm water, or two tablets per quart if the water is cold or cloudy. Each bottle of iodine tablets should have instructions for mixing and how long to wait before drinking the water. If no directions are available, follow the mixing instructions given for chlorine above.

Commercial purifying systems
A filter is not a purifying system. In general, filters remove protozoans; microfilters remove protozoans and bacteria; and purifiers remove protozoans, bacteria, and viruses. A purifier is simply a microfilter with iodine and carbon elements added. The iodine kills viruses, and the carbon element removes the iodine taste and reduces the presence of organic chemical contaminants such as pesticides, herbicides, and chlorine, as well as heavy metals. Unlike filters, purifiers must be registered with the EPA to demonstrate effectiveness against waterborne pathogens, including cysts, bacteria, and viruses. A purifier costs more than a filter but reduces your health risk as waterborne viruses become more and more prevalent. On the downside, a purifier tends to clog more quickly than most filters. Purifiers come in a pump style or a ready-to-drink bottle design.

Pump purifiers
Probably the best-known pump purifier is the PUR Explorer, an ideal system for the wilderness traveler. This is an easy-to-use, high-output, self-cleaning system that weighs 20 ounces. Its 0.3-micron filtering cartridge can produce up to 1.5 liters of water per minute and under normal conditions will not need to be replaced until it has provided around 100 gallons (400 liters) of drinkable water. Follow the manufacturer's guidelines for maintenance.

Bottle purifiers

Probably the best-known bottle purifier is the 34-ounce Exstream Mackenzie, also ideal for the wilderness traveler. The benefit of a bottle purifier is its ease of use—simply fill the bottle with water and start drinking, following the manufacturer's directions. It requires no assembly or extra space in your pack. On the downside, it filters only about 26 gallons (100 liters) before you'll need to replace the cartridge, and unless you carry several, you'll need to be in an area with multiple water sources throughout your travel. Maintenance of these systems is simple. When not in use, allow the cartridge to completely dry before storing; before resuming use, flush the system several times with tap water.

Purifier care

To increase the longevity of your system, use the following guidelines (in no way should they supersede the manufacturer's recommendations):

- *Cleaning.* Clean, scrub, and disinfect the filter after each use, following the manufacturer's guidelines. *Note:* Some filters should not be scrubbed, and some are self-cleaning.
- *Use clean water.* Whenever possible, procure clean water such as that found in a creek's pools or similar areas. To avoid sand, mud, and debris, keep the suction hose away from the water's bottom; this may require the use of a foam float.
- *Muddy water.* If muddy water is your only choice, fill a clean container with the muddy water and let it sit for several hours (overnight if time permits) or until the sediment has settled to the bottom of the container. Or you could first run the water through an improvised filter, as described above.
- *Backwash the filter.* Backwash the filter according to the manufacturer's recommended schedule. This process helps remove any accumulated debris from the system.

SURVIVAL TIPS

OTHER WATER YOU SHOULD AVOID
Avoid collecting water from areas with oily films or slicks, significant algae overgrowth, or near dead animals. If this is your only water source, you may decide to dig a natural filtration system in the ground close by as

a means of collecting this water, or perhaps pour it through a three-tiered filter and then treat it prior to drinking it.

DRINK ENOUGH WATER

If you are thirsty, you are already dehydrated. Don't wait until it is too late. On average, you should drink at least ½ quart of water an hour when temperatures are below 100 degrees F and 1 quart an hour when the temperature is higher than that.

9
Food

In a desert or survival situation, food should be deemphasized. When water is available, however, eating food will help sustain your energy level. The ideal diet contains these five basic elements:
1. Carbohydrates. Easily digested food elements that provide rapid energy. Most often found in fruits, vegetables, and whole grains.
2. Protein. Helps with the building of body cells. Most often found in fish, meat, poultry, and blood.
3. Fats. Slowly digested food elements that provide long-lasting energy, which is normally used once the carbohydrates are gone. Most often found in butter, cheese, oils, nuts, eggs, and animal fats.
4. Vitamins. Provide no calories but aid in the body's daily function and growth. Vitamins occur in most foods eaten, and if you maintain a well-balanced diet, you will rarely become depleted.
5. Minerals. Provide no calories but aid with building and repairing the skeletal system and regulating the body's normal growth. Like vitamins, minerals are obtained when a well-balanced diet is followed. They are also often present in water.

Even in harsh desert conditions, the ideal daily diet consists of approximately 50 to 70 percent carbohydrates, 20 to 30 percent proteins, and 20 to 30 percent fats. The time required to convert carbohydrates, proteins, and fats into simple sugars increases—in that order—due to the complexity of the molecule.

FOODS TO TAKE

If backpacking, weight will be an issue. Take dry foods like cereal, pasta, rice, wheat, and oatmeal, or purchase freeze-dried meals, which are a great option but tend to be expensive. If you prefer to carry enough water to drink but not enough to reconstitute food, consider using the military MREs (meals ready to eat), which don't require reconstitution.

As long as you have planned out the food for your trip properly and nothing goes wrong, you'll never need to look for food elsewhere. Should you find yourself short, however, or perhaps in a survival situation, you may need to look to Mother Nature to replenish your supply.

PLANTS

It has been said that more than three hundred thousand species of plants can be found on the earth's surface. With this in mind, it seems logical that plants can be a major source of your diet. The best way to learn whether a plant is edible is from those who are indigenous to the area, along with a good plant reference book. Still, be careful, and always positively identify a plant before eating it. If you don't have any references and need to establish the edibility of a plant, use the universal edibility test. This test should be used only under the most extreme conditions, when survival doesn't seem likely otherwise.

UNIVERSAL EDIBILITY TEST

General rules
1. Ensure that there's an abundant supply of the plant.
2. Use only fresh vegetation.
3. Always wash your plants with treated water.
4. Perform the test on only one plant or plant part at a time.
5. During the test, don't consume anything else other than purified water.
6. Don't eat eight hours prior to starting the test.

Avoid plants with these characteristics
These are general guidelines; there are exceptions.
1. Mushrooms or mushroomlike appearance.
2. Umbrella-shaped flower clusters (resembling those of Queen Anne's lace or dill).
3. Sap that is milky or turns black when exposed to the air.
4. Bulbs (resembling onions or garlic).

Six characteristics of plants to avoid

5. Carrotlike leaves, roots, or tubers.
6. Bean- or pealike appearance.
7. Shiny leaves or fine hairs.
8. Fungal infection (common in spoiled plants procured off the ground).

To test a plant
1. Break the plant into its basic components: leaves, stems, roots, buds, and flowers.
2. Test only one part of the potential food source at a time.

3. Smell the plant for strong or acidic odors. If present, it may be best to select another plant.
4. Prepare the plant part in the fashion in which you intend to consume it (raw, boiled, baked).
5. Place a piece of the plant part being tested on the inside of your wrist for fifteen minutes. Monitor for burning, stinging, or irritation. If any of these occur, discontinue the test, select another plant or another component of the plant, and start over.
6. If you experienced no reaction, hold a small portion, about a teaspoonful, to your lips and monitor for five minutes. If any burning or irritation occurs, discontinue the test, select another plant or another component of the plant, and start over.
7. Place the plant portion on your tongue, holding it there for fifteen minutes. Do not swallow any of the plant juices. If any burning or irritation occurs, discontinue the test, select another plant or another component of the plant, and start over.
8. Thoroughly chew the teaspoonful of the plant part for fifteen minutes. Do not swallow any of the plant or its juices. If you experience a reaction, discontinue the test, select another plant or another component of the plant, and start over. If there is no burning, stinging, or irritation, swallow the chewed plant material.
9. Wait eight hours. Monitor for cramps, nausea, vomiting, or other abdominal irritations. If you experience a reaction, induce vomiting and drink plenty of water. Discontinue the test, select another plant or another component of the plant, and start over.
10. If no problems are experienced, eat ½ cup of the plant, prepared in the same fashion as before. Wait another eight hours. If no ill effects occur, the plant part is edible when prepared in the same fashion as tested.
11. Test all parts of the plant you intend to use. Some plants have both edible and poisonous sections. Do not assume that a part that is edible when cooked is edible when raw, or vice versa. Always eat the plant in the same fashion in which the edibility test was performed on it.
12. After the plant is determined to be edible, eat it in moderation. Although considered safe, large amounts may cause cramps and diarrhea.

Aggregate berries are 99 percent eligible.

THE BERRY RULE
In general, the edibility of berries can be classified according to their color and composition. The following are approximate guidelines to help you determine whether a berry is poisonous. In no way should the berry rule replace the edibility test. Use it as a general guide to determine whether the edibility test needs to be performed upon the berry. The only berries that should be eaten without testing are those that you can positively identify as nonpoisonous.
1. Green, yellow, and white berries are 10 percent edible.
2. Red berries are 50 percent edible.
3. Purple, blue, and black berries are 90 percent edible.
4. Aggregate berries, such as thimbleberries, raspberries, and blackberries, are considered 99 percent edible.

EDIBLE PARTS OF A PLANT
Some plants are completely edible, whereas others have both edible and poisonous parts. *Unless you have performed the edibility test on the whole plant, eat only the parts that you know are edible.* Plants can be broken

down into several distinct components: underground, stems and leaves, flowers, fruits, nuts, seeds and grains, gums, resins, and saps.

Underground (tubers, roots and rootstalks, and bulbs)
Found underground, these plant parts have a high degree of starch and are best served baked or boiled. Some examples are potatoes (tuber), cattail (root and rootstalk), and wild onions (bulbs).

Stems and leaves (shoots and stems, leaves, and cambium)
Plants that produce stems and leaves are probably the most abundant source of edible vegetation in the world. Their high vitamin content makes them a valuable component to our daily diet. *Shoots* grow like asparagus and are best when parboiled: boiled five minutes, drained, and boiled again until done. Cattail is an example of a shoot you may find around a desert spring. *Leaves* may be eaten raw or cooked, but to achieve the highest nutritional value, they are best eaten raw. Flat cacti leaves are an example of edible leaves. *Cambium* is the inner bark found between the bark and the wood of a tree. It can be eaten raw, cooked, or dried and then pulverized into flour.

Flowers (flowers, buds, and pollen)
Flowers, buds, and pollen are high in food value and are best when eaten raw or in a salad. Some examples are abel (flower), rosehips (buds), and cattail (pollen).

Fruits (sweet and nonsweet)
Fruits are the seed-bearing part of a plant and can be found in all areas of the world. They are best when eaten raw so they retain all of their nutritional value, but they may also be cooked. Examples of desert fruits are prickly pears and dates.

Nuts
Nuts are high in fat and protein and can be found around the world. Most can be eaten raw, but some, like acorns, require leaching with several changes of water to remove their tannic acid content.

Seeds and grains

Seeds and grains are a valuable food resource that should not be overlooked. Some examples are grasses and millet. These are best eaten when ground into flour or roasted. Black or purple grass seeds often contain a fungal contamination that can make you sick and should not be eaten. Saint John's bread is a seed that can be found in the desert.

Gums and resins

Gums and resins are saps that collect on the outside of trees and plants. Their high nutritional value makes them a great augment to any meal. They can be found on pine and maple trees, for example.

DESERT PLANTS

Hot, dry deserts, such as the Sahara and Middle Eastern Deserts, have little vegetation, whereas cooler, wetter deserts, such as the Great Basin and Gobi Deserts, are often relatively productive. Plants derive food from sunlight, and thus desert plants rarely have a starvation problem. Obtaining water, on the other hand, can be difficult for desert plants, and the presence or lack of desert vegetation is a direct result of water availability.

Survival strategies

In order to survive on limited water, the three basic desert plants employ different strategies.

Annual or ephemeral plants

Annual or ephemeral plants take root and grow when moisture is plentiful following a rainfall, and they die or become dormant when the drought returns. The dormant seeds are heat- and drought-resistant and remain in the soil until the next year's annual rains.

Phreatophytes

Phreatophytes, a term meaning water-loving plants, often have taproots extending deep enough to reach the water table. Mesquite is a phreatophyte with a deep root system that is an adaptation to the hot desert.

Xerophytes

Xerophytic plants, such as sagebrush, mesquite, yucca, and saguaro and prickly pear cactus, have the ability to extract water from extremely dry soil and adapt to decrease water vapor loss from the leaves. Leaves may possess surface hairs to reflect light and to slow wind flow; have a reduced surface area; roll or curl under hot, dry conditions; or drop during extreme drought. Succulents such as cactus have a highly efficient water-gathering root system and the ability to store water in a spongy tissue at the center of the stem or root.

Edible desert plants

Each of the numerous deserts around the world has edible vegetation unique to it. Although food is a lower priority than water, you should still become familiar with edible plants indigenous to your area of travel before departing. A few plant food sources common in most deserts include the following.

Prickly pear cactus

There are many species of prickly pear, but most are characterized by flat, fleshy, oval-shaped pads covered with spines. Most produce yellow, red, or purple flowers. Both the fruit and pads are edible. Before eating the pads, you should use a fire or hot coals to scorch the spines off and soften the outer skin. If you cannot build a fire, then carefully attempt to peel the skin away with a knife.

Grasses

Grasses can often be found in meadows, drainages, and dry riverbeds. The stems, roots, and leaves may be eaten raw or cooked, and grass can be boiled in water to make a good broth or tea. Do not eat black or purple grass seeds, which indicate a fungal contamination that, if eaten, could cause severe illness or death.

Sotol

Sotol can be found on rocky slopes in many desert grasslands. The plant has hundreds of 3-foot-long ribbonlike leaves that shoot up from a central rounded ball core. Although similar in appearance to yucca, sotol's light

Food   125

Prickly pear cacti

green leaves have small teeth along the sides. Each year in early summer, the sotol plant produces a flower stalk that can grow up to 12 feet tall and has thousands of greenish white flowers that grow in a dense cluster. The rounded ball's heart is the edible part of sotol. It can be procured using a digging stick or similar device. To prepare, cook in a rock-lined baking pit until it no longer tastes bitter, which usually takes several days. The leaf bases can be eaten in the same manner as an artichoke. The stalks can also be used for building material and the leaves for making baskets, mats, and cordage.

Century plant
Similar to sotol, century plant is often found on rocky slopes in many desert grasslands. The plant has spine-tipped leaves that are about 18 inches long and 3 inches wide and crop up from a woody heart, and it produces a flower stalk that can reach up to 15 feet high. The heart of the century plant is

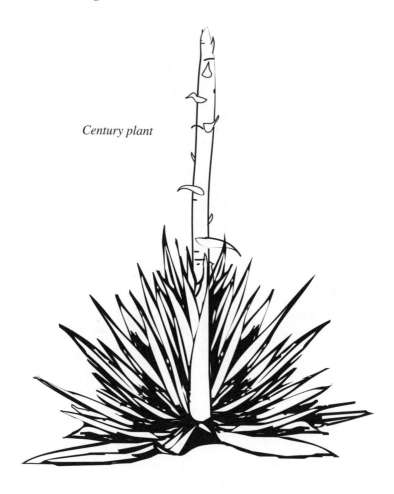
Century plant

edible and should be prepared in the same fashion as sotol. The stalks can also be used for building material and the leaves for making baskets, mats, and cordage.

Yucca
Like the century plant and sotol, yucca is found on rocky slopes in many deserts. There are many species of yucca, most with stiff-pointed swordlike leaves and towering stalks with creamy white, waxy flowers. The plant may

have one or several stalks that range in height from 3 to 10 feet tall. Yucca flower petals can be eaten raw or cooked, or dried for later consumption. The fruit can be boiled until very soft, peeled, and seeded before eaten. The stalks can be used for building material and the leaves for making baskets, mats, and cordage. In addition, the roots of the soaptree yucca plant have a high content of saponins, a soaplike compound, making it a favorite shampoo and soap for indigenous peoples.

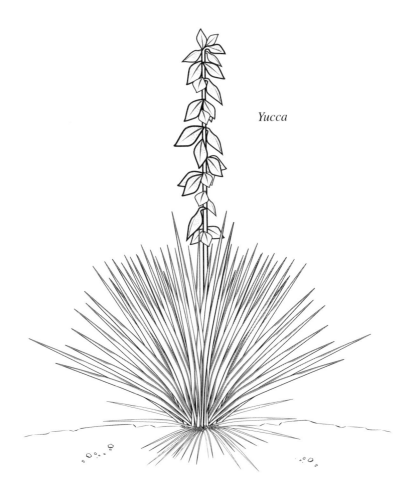

Yucca

128   Surviving the Desert

Desert amaranth
Amaranth grows in most desert climates. Although there are many species of amaranth, most are short-lived annual herbs with green leaves and thick, erect fleshy stems. The young shoots and leaves can be eaten raw or cooked, or dried for later use. After removing the chaff from the seeds, they can be cooked like popcorn or ground into flour for use in breads.

Date palm
The date palm is found in the deserts of North Africa and the Middle East. This tall tree has deep roots and is supported by a trunk growing from the woody leaf base. The feather-shaped leaves are about 15 feet long and support clusters of flowers or dates. The ripened reddish brown date provides an excellent and abundant nutritious fruit. The branches of the date palm can be used for shade and roofing materials and to improvise con-

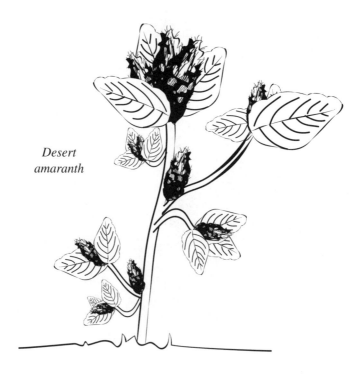

Desert amaranth

Date palm

tainers. Date trees are a source of food for the inhabitants of the Arabian Peninsula, and their leaves provide shade from the intense desert sun.

Acorns
Acorns are found on many oaks and are edible when leached. Acorns collected from the tree are less prone to have insect, heat, or drying damage than those on the ground. An acorn is ready to pick when it can easily separate from its cap without tearing the seed coat. If you pick a green acorn, keep it—it will brown quickly. Leaching is necessary to remove the tannins from the acorn. To properly leach acorns, remove the meat from the shell (this may require boiling it for ten to fifteen minutes in order to soften the shell), and place the meat in a container. Cover the shelled acorns with water, bring the water to a boil, and drain. Repeat the process three or four times, or until the water is clear and the acorns are no longer bitter. If you can't boil water but a stream is close, place the acorns in a

Acorn

porous material and let them soak in a running stream for one to two days. Acorns can be eaten raw or dried, or ground into flour for use in pancakes or bread.

Cattails

Although cattails are found in moist, swamplike areas, you may discover them in the desert around a natural spring or some other underground water source. Cattails provide several edible portions, including the roots, sprouts, shoots, flower spikes, seeds, and pollen. Cattails are easy to recognize when the stalks are topped with the dense, brown, oblong-shaped seed clusters that appear after the flowers have fallen off. Long, slender, sword-shaped leaves branch off the stalks that can reach up to 8 feet high. Cattail roots are best from late fall through early spring, when they have a high starch concentration, and can be eaten raw or cooked. Sprouts grow from the roots and can be gathered from late summer to winter. They are often cooked like potatoes. The green shoots are best when gathered during the spring before they reach 2 feet in height. Peel away the shoot's outer layer until you reach the white tender core, and eat it raw or steamed. The immature green flower spikes are gathered during late spring before they begin to produce pollen. To prepare, husk and cook the shoot like corn. Seeds are found in the lower section of the pod and are present during the summer. These high-protein seeds can be mashed into flour and used in any number of recipes. Pollen is found on the upper section of the pod and is present during early summer. To procure the yellow pollen, rub, shake, or strip it off into a container or bag. This yellow powder is very high in protein and can be eaten raw, cooked as a cereal, or used with flour.

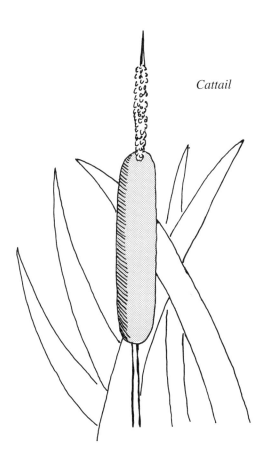

Cattail

BUGS

Many cultures around the world eat bugs as part of their routine diet. Pan-fried locusts are considered a delicacy in Algeria and several Mexican states. In Malaysia, bee larvae are considered a special treat. Our phobia about eating bugs is unfortunate, as they contain ample amounts of protein, fats, carbohydrates, calcium, and iron. Compared with cattle, sheep, pigs, and chickens, bugs are far more cost effective to raise and have far fewer harmful effects related to their rearing. Although bugs are not harvested for food in the United States, those of us who purchase our foods at the store are eating them every day. The FDA allows certain levels of bugs to be present in various foods. The accepted standards are for up to sixty aphids in 3½ ounces of broccoli, two or three fruit fly maggots in 200 grams of tomato juice, one hundred insect fragments in 25 grams of curry powder, seventy-four mites in 100 grams of canned mushrooms, thirteen insect heads in 100 grams of fig paste, and thirty-four fruit fly eggs in a cup of raisins.

A study done by Jared Ostrem and John VanDyk of the Entomology Department at Iowa State University, comparing the nutritional value of various bugs to that of lean ground beef and fish, showed the following results per 100 grams:

	protein (g)	fats (g)	carbohydrates (g)	calcium (mg)	iron (mg)
crickets	12.9	5.5	5.1	75.8	9.5
small grasshoppers	20.6	6.1	3.9	35.2	5.0
giant water beetles	19.8	8.3	2.1	43.5	13.6
red ants	13.9	3.5	2.9	47.8	5.7
silkworm pupae	9.6	5.6	2.3	41.7	1.8
termites	14.2	N/A	N/A	0.05	35.5
weevils	6.7	N/A	N/A	0.186	13.1
lean ground beef (baked)	24.0	18.3	0	9.0	2.09
fish (broiled cod)	22.95	0.86	0	0.031	1.0

Bugs are a good source of food.

Bugs can be found throughout the world, and they are easy to procure. In addition, the larvae and grubs of many are edible and can be easily found in rotten logs, underground, or under the bark of dead trees. Although a fair number of bugs can be eaten raw, it's best to cook them in order to avoid the ingestion of unwanted parasites. As a general rule, avoid bugs that carry disease (flies, mosquitoes, and ticks), poisonous insects (centipedes and spiders), and bugs that have fine hair, bright colors, and eight or more legs.

CRUSTACEANS

Crustaceans such as saltwater shrimp, crayfish, and lobster might be found in the waters off a coastal desert.

SALTWATER SHRIMP

Saltwater shrimp live on or near the sea bottom. Since these shrimp are attracted to light, it's best to hunt them during a full moon or lure them to the water's surface with a flashlight. Once spotted, simply scoop them up with a net or pluck from the water with your hand.

SALTWATER CRAYFISH AND LOBSTER

Saltwater crayfish and lobsters are found on the ocean bottoms in 10 to 30 feet of water. To catch during the day, use a lobster trap or baited hook.

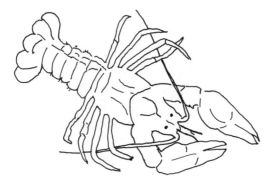

Crustaceans may be found near coastal deserts.

Improvise a lobster trap by securing bait to the inside bottom of a container the size of a large coffee can. If using a can, puncture small holes in the bottom so that water can pass through it. Attach enough line to the trap's sides that it can be lowered to the stream's bottom. Once the trap is in place, it won't take long before a lobster or crayfish crawls inside to eat the bait, so check it often. When pulling the container from the water, do it swiftly but with enough control to avoid pouring out your dinner.

MOLLUSKS

Mollusks might be found in waters off a coastal desert. However, they should be avoided from April to October. During this time, waters are prone to red tides that contain a poison harmful to humans, and mollusks are known to accumulate this toxin. Also avoid shellfish that are not covered by water at high tide. Saltwater mollusks include bivalves, such as clams, oysters, scallops, and mussels, as well as limpets and chitons. All can be boiled, steamed, or baked.

Avoid mollusks that are not covered at high tide.

REPTILES

LIZARDS

Lizards are adaptive creatures and can be found in most deserts. Only two lizards are poisonous: The Gila monster and the Mexican beaded lizard, both found in the U.S. Southwest, Mexico, and Central America, though they are slow moving and docile, are poisonous and should be avoided. All lizards are edible but should be skinned and gutted prior to consuming. They can be caught with a noose on the end of a stick, or stunned and killed with a club or rock. Broil or fry the meat.

SNAKES

All snakes—poisonous and nonpoisonous—are edible. Many kinds of snakes inhabit desert areas and can be located almost anywhere there is cover. For best results, hunt for them in the early morning or evening hours. To catch or kill a snake, first stun it with a thrown rock or stick, then use the forked end of a long stick to pin its head to the ground. Kill it with a rock, knife, or another stick. Be careful throughout this procedure, especially when dealing with poisonous snakes. Snakes can be cooked in any fashion, but all should be skinned and gutted. To skin a snake, sever its head (avoid

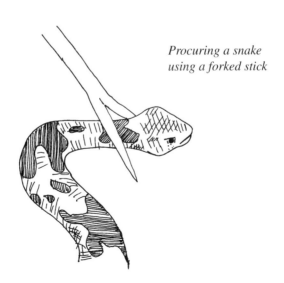

Procuring a snake using a forked stick

accidental poisoning by burying the head) and peel back the skin until you can grab it and pull it down, inside out, the length of the snake. If you can't pull it free, make a cut down the length of the snake to help free the skin. The entrails will usually come out during this process; if not, grab them at the top and pull them down to remove them.

FISH

Fish are commonly found in almost all sources of water and may provide an excellent source of food in a coastal desert. The best times to fish are just before dawn or just after dusk, at night when the moon is full, and when bad weather is imminent. You can procure fish with tackle, gill or scoop nets, spears, fish traps, or even your bare hands. Avoid saltwater fish during red tide (from April to October) and any fish that has a slimy body, bad odor, suspicious color (gills should be pink and scales pronounced), or flesh that remains indented after being pressed on.

FISHING TACKLE

If you have fishing tackle, use it. If you don't, you'll need to improvise. Crude tackle isn't very useful for catching small fish like trout, but it can be effective with larger fish such as carp, catfish, or whitefish.

Hooks

Some commonly used improvised hooks are skewer and shank hooks, made from bone, wood, or plastic, and safety pin hooks.

Skewer hook

A skewer hook is a sliver of wood, bone, or plastic that is notched and tied at the middle. When baited, this hook is turned parallel to the line, making it easier for the fish to swallow. Once the fish takes the bait, a simple tug on the line will turn the skewer sideways, lodging it in the fish's mouth.

Shank hook

A shank hook is made by carving a piece of wood, bone, or plastic into the shape of a hook. It should be notched and tied at the top. When the fish swallows the hook, a gentle tug on the line will set it by causing the hook end to lodge in the throat.

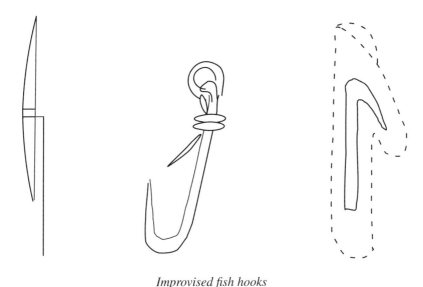

Improvised fish hooks

Safety pin hook

A safety pin can be manipulated to create a hook, as shown in the illustration. Depending on the size of the safety pin, this hook can catch fish of various sizes and is a good option.

Lines

If you don't have fishing line, use a 10-foot section of improvised cordage (see back of book). Although you could attach your line to a single pole, I'd advise setting out multiple lines tied to the end of one or several long, straight branches. Sticking these poles into the ground allows you to catch fish while attending to other chores. The goal is to return and find a fish attached to the end of each line.

To attach a standard hook, safety pin, or fixed loop to your line, use an improved clinch knot. All other improvised hooks can be attached to line using any knot. Following are the steps to attach a hook with a clinch knot.

1. Run the free end of the line through the hook's eye and fold it back onto itself.

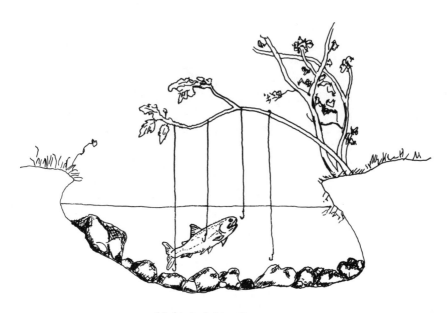

Multiple fishing lines set out

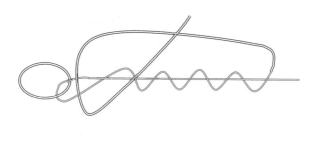

Improved clinch knot

2. Wrap the free end up and around the line six or seven times.
3. Run the line's free end down and through the newly formed loop that is just above the hook's eye.
4. Finally, run the line through the loop formed between the two lines twisted together and the free end that just went though the loop next to the hook's eye.
5. Moisten the knot and pull it tight. Cut the excess line.

GILL NET

If you have the time and materials to construct a gill net, it's worth doing. It's a very effective method of procuring fish with limited work once the construction is complete, and it will work for you while you attend to other needs. If you have parachute cord or a similar material, its inner core provides an ideal material for making a net. If not, use braided cordage (details on improvising cordage appear at the back of the book). In order for the net to stay clear of debris, it should be placed at a slight angle to the current, using stones to anchor the bottom and wood to help the top float. Follow these steps to make a gill net:

1. Tie a piece of line between two trees at eye height. The bigger the net you want, the farther apart the trees should be.
2. Using a girth hitch, tie the center of each piece of inner core line or other material to the upper cord. Use an even number of lines. Space the lines apart at the width you desire for your net's mesh. For creeks and small rivers, 1 inch is about right.
3. Starting at either end, skipping the line closest to the tree, tie the second and third lines together with an overhand knot. Continue on down the line, tying together the fourth and fifth, sixth and seventh, and so on. When you reach the end, there should be one line left. If you are concerned about the mesh size, first tie a guideline between the two trees. For a 1-inch mesh, tie this line 1 inch below the top line, and use it to determine where the overhand knots should be placed. Once a row of knots is completed, move this guideline down another inch.
4. Moving in the opposite direction, tie the first line to the second, third to the fourth, and so on. When you reach the end, there shouldn't be any lines left.
5. Repeat the previous two steps until done.

140   Surviving the Desert

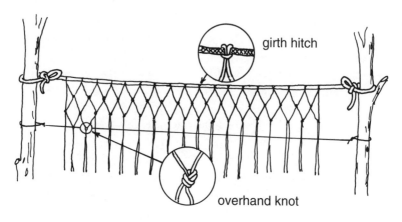

Constructing a gill net

6. When done, run parachute line or other material along the net's sides and bottom to help stabilize it.

SCOOP NET

A scoop net can help secure a line-caught fish or can be used alone to scoop a fish out of the water. To make a scoop net, bend the two ends of a 6-foot sapling or similar material together to form a circle, allowing some extra length for a handle. You can also form a circle with the ends of a forked branch. Lash the ends together. The net's mesh can be made using the same method described for building a gill net, tying the initial girth hitch to the sapling. Once the net is the appropriate size, tie all the lines together with an overhand knot, and trim off any excess. A scoop net should be used in shallow water or other area where fish are visible. Because you'll need to compensate for light refraction below the water, first place the net into the water to obtain the proper alignment. Next, slowly move the net as close to the fish as possible, and allow them to become accustomed to it. When ready, scoop the fish up and out of the water.

SPEAR

A spear can be used to procure both fish and small mammals. To make a straight spear, sharpen one end of a long, straight sapling to a barbed point. If practical, fire-harden the tip to make it more durable by holding it a few inches above a hot bed of coals until brown. To make a forked spear, fire-harden the tip of a long, straight sapling. Snugly lash a line around the stick 6 to 8 inches down from one end. Using a knife, split the wood down the center to the lash. To keep the two halves apart, lash a small wedge between them. (For best results, secure the wedge as far down the shaft as possible.) Sharpen the two prongs into inward-pointing barbs.

Using a spear to procure fish is a time-consuming challenge, but under the right circumstances it can yield a tasty supper. You'll need to compensate for light refraction below the water's surface. To obtain proper alignment, place the spear tip into the water before aiming. Moving the spear tip slowly will allow the fish to get accustomed to it until you are ready. Once you have speared a fish, hold it down against the bottom of the stream until you can get your hand between it and the tip of the spear.

Forked wooden spear

FISH TRAP

Fish traps would perhaps be better called corrals, since the idea is to herd the fish into the fenced enclosure.

When building these traps in ocean water, select your location during high tide and construct the trap during low tide. On rocky shores, use natural rock pools; on coral islands, use the natural pools that form on the reefs; and on sandy shores, create a dam on the lee side of the offshore sandbar. If you can, block all the openings before the tide recedes. Once the tide goes back out, you can use either a scoop net or spear to bring your dinner ashore.

PREPARING FISH

To prevent spoilage, prepare the fish as soon as possible. Always do this well away from your shelter. Gut the fish by cutting up its abdomen and then removing the intestines and large blood vessels that lie next to the backbone. Remove the gills and, when applicable, scale and/or skin the fish.

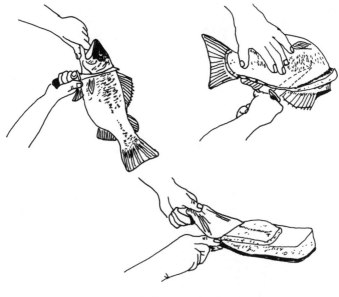

Filleting a fish

On bigger fish, you may want to fillet the meat off the bone. To do this, cut behind the fish's gill plates on each side of its head, and slide the knife under the meat next to the backbone. Keeping the knife firmly placed against the backbone, begin slicing toward the tail. Then, holding the tail's skin, slide the knife between the skin and the meat, cutting forward with a slight sawing motion.

BIRDS

Almost all birds are edible, and in the desert, most can be found close to a water source. If nests are near, eggs may also be available for consumption. Eggs are easy to obtain, and young birds are easy to procure with snares, with a baited hook, or on occasion, by clubbing. The Ojibwa bird snare is a simple snare used to catch a bird.

OJIBWA BIRD SNARE

The Ojibwa bird snare is effective, but it requires time and materials to create. If you have both, it may be worthwhile to set one out. Find a sapling or similar material that is 1 to 2 inches in diameter, and cut off the top so it's 4 or 5 feet high. To prevent birds from landing on the top, carve it into a point. The bait or a shiny object also can be attached here. Make a hole slightly larger than ¼ inch in diameter near the top of the sapling. To make a perch, cut a ¼-inch-diameter stick 6 to 8 inches long. If you prefer, you can sharpen one end of the stick and attach the bait there.

Make a slip knot or noose at one end of a piece of 3- to 4-foot line. The noose should be about 4 inches in diameter. Tie an overhand knot 1 to 2 inches beyond the noose. This knot is instrumental in securing the perch to the sapling until a bird lands on it. Insert the free end of the snare line through the hole in the sapling, and pull until the knot reaches the opening. Insert the perch into the hole, and use the knot to lightly secure it in place. If using the perch to hold the bait, attach the bait first.

Tie a rock or heavy stick to the free end of the line. It must be heavy enough to pull the noose tight once the bird dislodges the perch from the sapling. Lay the noose on top of the perch. It may be easier to tie the rock to the line before inserting the perch into the hole.

An alternative to the rock would be to cut a 3-foot-long sapling that is then attached to the back of the Ojibwa's pole, with its top end below the

144 Surviving the Desert

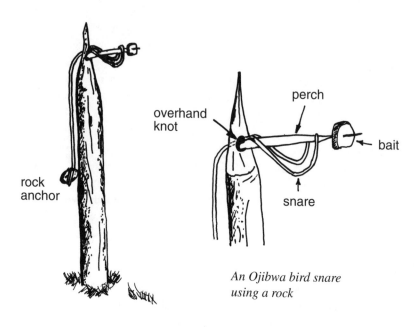

An Ojibwa bird snare using a rock

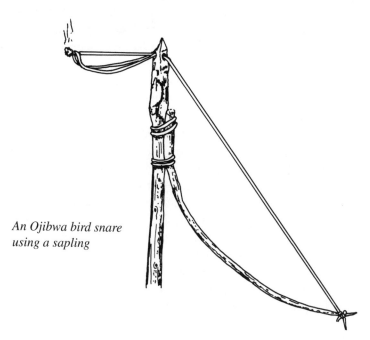

An Ojibwa bird snare using a sapling

perch's hole. Carve a small notch on the underside of the twig so that the free end of the snare line can be attached. Finally, adjust the line's length so that when the snare is armed, enough tension is created to quickly tighten the noose when the bird dislodges the perch from the sapling.

PREPARING BIRDS
Pluck all birds unless they are scavengers or seabirds, which should be skinned. Leaving the skin on other kinds of birds will retain more of their nutrients when cooked. Cut the neck off close to the body. Cut open the chest and abdominal cavity, and remove the insides. Save the neck, liver, heart, and gizzard, which are all edible. Before eating the gizzard, split it open and remove the stones and partially digested food. Cook in any desired fashion. Cook scavenger birds a minimum of twenty minutes to kill parasites.

MAMMALS
Mammals provide a great source of meat. Signs that indicate their presence include well-traveled trails, usually leading to feeding, watering, or bedding areas; fresh tracks and droppings; and fresh bedding areas, such as nests, burrows, or trampled-down field grass. The odds of obtaining a big-game animal without a rifle are small, and the risk of injury from trying to snare one is too high. Small game can be procured using handheld devices or by setting out snares.

HANDHELD WEAPONS
Handheld weapons include rocks, throwing sticks, spears, bolas, weighted clubs, slingshots, and rodent skewers. Skill and precise aim are the keys to success when using these devices, and acquiring them requires practice.

Rocks
Hand-size stones can be used to stun an animal long enough for you to approach and kill it. Aiming toward the animal's head and shoulders, throw the rock as you would a baseball.

Throwing stick
The ideal throwing stick is 2 to 3 feet long and thicker or weighted on one end. Holding the thin or lighter end of the stick, throw it in either an

overhand or sidearm fashion. For best results, aim for the animal's head and shoulder.

Spear

For details on how to construct a spear, refer to the section on fish procurement, above. A throwing spear should be 5 to 6 feet long. To throw a spear, hold it in your right hand (if left-handed, reverse these instructions), and raise it above your shoulder so that the spear is parallel to the ground. Position your hand at the spear's center point of balance. Place your body so that your left foot is forward and your trunk is perpendicular to the target. Point your left arm and hand toward the animal to help guide you when throwing the spear. Once positioned, thrust your right arm forward, releasing the spear at the moment that will best enable you to strike the animal in the chest or heart.

Bola

A bola is a throwing device that immobilizes small game long enough for you to approach and kill it. To construct a bola, use an overhand knot to tie three 2-foot-long lines together about 3 to 6 inches from one end. Securely attach a ½-pound rock to the other end of each of the three lines. To use the bola, hold the knot in your hand, and twirl the lines and rocks above your head or out to your side until you have attained adequate control and speed.

Bola

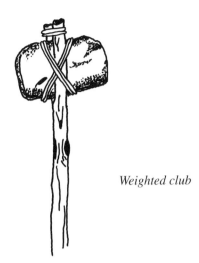

Weighted club

Once this is accomplished, release your grip when the bola is directed toward the intended target.

Weighted club
A weighted club can not only be used to kill an animal at close range, but it's also a valuable tool for meeting other survival needs. To construct one, find a rock that is 6 to 8 inches long, 3 to 4 inches wide, and about 1 inch thick. Cut a 2- to 3-foot branch of straight-grained wood that is 1 to 2 inches in diameter. Hardwood is best, but softwood also works. Snugly lash a line around the stick 6 to 8 inches down from one end. Split the wood down the center and to the lash with a knife. Insert the stone between the wood and as close to the lashing as possible, and secure the rock to the stick with a tight lashing above, below, and across the rock. You can also use a strong forked branch and secure the rock between the two forked branches. Use the weighted club in the same fashion as a throwing stick.

Slingshot
A slingshot is a fairly effective tool for killing small animals. To construct one, you'll need elastic cord, bungee cord, or surgical tubing, as well as webbing or leather to make a pouch. Cut a strong forked branch with a base 6 to 8 inches long and 3- to 5-inch forked sides. Carve a notch around the top of each forked side, ½ inch down from the top. Cut two 10- to 12-inch

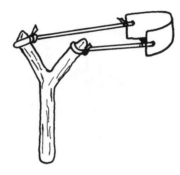

Improvised slingshots are effective weapons for procuring small game.

pieces of elastic cord or line, and secure them to the branches by wrapping one end of each cord around the carved notches, and then tightly lashing them in place. Cut a piece of webbing or leather 3 inches long and 1 to 2 inches wide. Make a small hole that is centered and ½ inch in from each side. Using the free end of the two elastic cords, run ½ inch through the hole in the webbing or leather. Secure each cord to the webbing or leather by lashing it in place. To use the slingshot, hold a marble-size rock in the slingshot's pouch between the thumb and pointer finger of your right hand. Place your body so that your left foot is forward and your trunk is perpendicular to the target. Holding the slingshot with a straightened left arm, draw the pouch back toward your right eye. Position the animal between the forked branches, and aim for the head and shoulder region. Release the rock.

Rodent skewer
A forked spear made from a long sapling can be used as a rodent skewer. To use it, thrust the pointed end into an animal hole until you feel the animal. Twist the stick so that it gets tightly snagged in the animal's fur, then

Rodent skewer

pull the animal out of the hole. The rodent will try to bite and scratch you if it can, so keep it at a distance. Use a club or rock to kill it.

SNARES AND TRAPS

You can also procure game with snares or traps. Once placed, they continue to work while you can tend to other needs. It shouldn't take much to find the indigenous animals' superhighways. These trails are located in heavy cover or undergrowth, or parallel to roads and open areas, and most animals routinely use the same pathways. Although several snares are covered in this section, for squirrel- and rabbit-size game, a simple loop snare is the best method of procurement in all climates.

Simple loop snare

An animal caught in a simple loop snare will either strangle itself or be held secure until your arrival. To construct this type of snare, use either snare wire or improvised line that's strong enough to hold the mammal you intend to catch (details on how to make cordage appear at back of book). If using snare wire, start by making a fixed loop at one end. To do

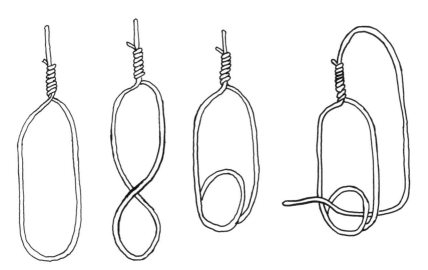

Four steps for constructing a simple loop snare

this, bend the wire 2 inches from the end, fold it back on itself, and twist or wrap the end of the wire and its body together, leaving a small loop. Twist the fixed loop at its midpoint until it forms a figure eight. Fold the top half of the figure eight down onto the lower half. Run the free end of the wire through the fixed loop. The size of the snare will determine the resultant circle's diameter. It should be slightly larger than the head size of the animal you intend to catch.

If using improvised line, make a slipknot that tightens down when the animal puts its head through it and lunges forward.

Avoid removing the bark from any natural material used in the snare's construction. If the bark is removed, camouflage the exposed wood by rubbing dirt on it. Since animals avoid humans, it's important to remove your scent from the snare. One method of hiding your scent is to hold the snaring material over smoke or underwater for several minutes prior to its final placement. Place multiple simple loop snares, at least fifteen for every animal you want to catch, at den openings or well-traveled trails so that the loop is at the same height as the animal's head. When placing a snare, avoid disturbing the area as much as possible. If establishing a snare on a well-traveled trail, try to use the natural funneling of any surrounding vegetation. If natural funneling isn't available, create your own with strategically placed sticks. (Again, hide your scent.) Attach the free end of the snare to a branch, rock, or drag stick, a big stick that either is too heavy for the animal to drag or will get stuck in the surrounding debris when the animal tries to move. Check your snares at dawn and dusk. Always make sure any caught game is dead before getting too close.

Slipknot

Simple loop snare

Funneling

Squirrel pole

A squirrel pole is an efficient means by which to catch multiple squirrels with minimal time, effort, or materials. Attach several simple loop snares to a 6-foot-long pole, then lean the pole onto an area with multiple squirrel feeding signs; look for mounds of pinecone scales, usually on a stump or a fallen tree. The squirrel will inevitably use the pole to try to get to his favorite feeding site.

Twitch-up strangle snare

An animal caught in a twitch-up strangle snare will either strangle itself or be held securely until your arrival. The advantage of the twitch-up snare over the simple loop snare is that it will hold your catch beyond the reach of other predatory animals that might wander by. To construct this type of snare, begin by making a simple loop snare out of either snare wire or strong improvised line. Find a sapling that, when bent to 90 degrees, is directly over the snare site you have selected.

Squirrel pole

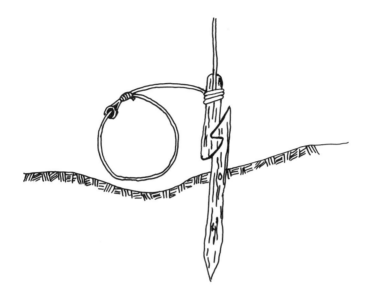

Two-pin toggle trigger

You'll need to construct a two-pin toggle trigger to attach the sapling to the snare while holding its tension. Procure two small forked or hooked branches that ideally fit together when the hooks are placed in opposing positions. If unable to find such branches, construct them by carving notches into two small pieces of wood until they fit together.

To assemble the twitch-up snare, firmly secure one branch of the trigger into the ground so that the fork is pointing down. Attach the snare to the second forked branch, which is also tied to the sapling at the location that places it directly over the snare when bent 90 degrees. To arm the snare, bend the twig and attach the two-pin toggle together. The resultant tension will hold it in place. Adjust the snare height to the approximate position of the animal's head. When an animal places its head through the snare and trips the trigger, it will be snapped upward and strangled by the snare. If you're using improvised snare line, it may be necessary to place two small sticks into the ground to hold the snare open and in a proper place on the trail.

154   Surviving the Desert

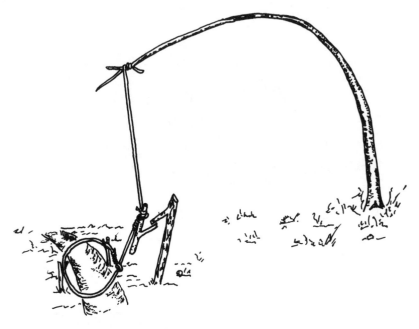

Twitch-up strangle snare using a two-pin toggle trigger

Figure-four mangle snare

A figure-four mangle snare is often used to procure small rodents such as mice, squirrels, and marmots. An animal caught in this snare will be mangled and killed. To construct this snare, procure two sticks that are 12 to 18 inches long and ¾ to 1 inch in diameter (the upright and diagonal pieces) and one stick that is the same diameter but 3 to 6 inches longer (the trigger).

Upright piece

Prepare the upright stick by cutting a 45-degree angle at its top end and creating a squared notch 3 to 4 inches up from the bottom. For best results, cut a diagonal taper from the bottom of the squared notch to the stick's bottom. This will aid in the trigger's release from the upright. In addition to being at opposite ends, the squared notch and the 45-degree angle must be perpendicular to one another.

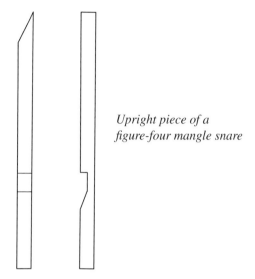

Upright piece of a figure-four mangle snare

Diagonal piece

To create the diagonal piece, cut a diagonal notch 2 inches from one end and a 45-degree angle on the opposite end. In addition to being at opposite ends, the diagonal notch and the 45-degree angle must be on the same sides of the stick.

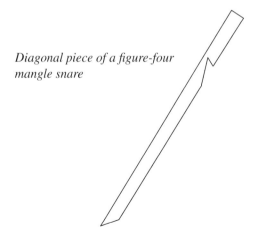

Diagonal piece of a figure-four mangle snare

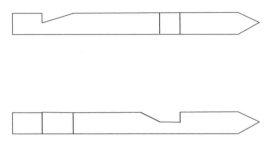

Figure-four trigger piece

Trigger

The trigger piece needs to have a diagonal notch cut 1 to 2 inches from one end and a squared notch created at the spot where this piece crosses the upright when the three sticks are put together. To determine this location, place the upright perpendicular to the ground, and insert its diagonal cut into the notch of the diagonal piece. Put the angled cut of the diagonal stick into the trigger's notch and hold it so that the number four is created between the three sticks when the trigger passes the upright's square notch. Mark the trigger stick and make a squared notch that has a slight diagonal taper from its bottom toward its other notched end. If you intend to bait the trigger, sharpen its free end to a point.

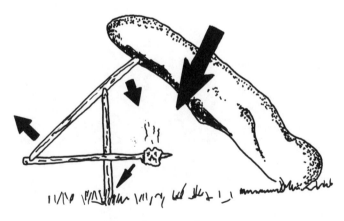

Figure-four mangle snare

Using the figure-four

To use a figure-four, put the three pieces together, and lean a large rock or other weight against the diagonal, at an approximate 45-degree angle to the upright. The entire structure is held in place by the tension between the weight and the sticks. This object will fall and mangle an animal that trips the trigger.

Paiute deadfall mangle snare

Another option is the Paiute deadfall mangle snare. Its touchy trigger system is a unique part of its design. To construct a Paiute snare, gather four slender branches and a short piece of line. This trap has five parts: upright, diagonal, trigger, bait stick, and line.

Upright

The upright, the only piece in contact with the ground, needs to have a flat bottom and beveled top. It should be long enough to create a 45-degree angle between your mangle device (most often a rock) and the ground.

Diagonal

The diagonal piece is approximately two-thirds the length of the upright. Prepare this piece by cutting a notch on one end, about 1 inch from the tip, and a circular groove around the other end, ½ inch up.

Trigger

The trigger is a small branch long enough to extend 1 inch beyond both sides of the upright when placed perpendicular to it.

Bait stick

The bait stick should be long enough to touch both the trigger stick (when in appropriate position) and the rock when it is held parallel to the ground.

Line

The line or cordage is attached to the diagonal piece's circular notch and needs to be long enough to wrap around the lower end of the upright while attached to the trigger. Attach the line to the trigger so that it's on the side that ends up opposite the line coming off the diagonal piece. It should be

158 Surviving the Desert

Paiute deadfall mangle snare

cut so that when the trap is set, it creates a 45-degree angle between the upper end of the diagonal and the upright piece.

Using the Paiute snare
Arm the Paiute snare as follows:
1. Tie the line to the circular groove of the diagonal stick.
2. Place the diagonal branch, with the notch side up, on the up side of the rock, forming a 45-degree angle between the rock and the ground.
3. Put the beveled side of the upright into the notch while maintaining the 45-degree angle. The upright should be placed so it is approximately perpendicular to the ground.
4. Attach the line to the trigger.
5. Run the line (off of the diagonal branch's groove) around the upright so that the trigger is perpendicular to the upright and on the side away from the rock.
6. Hold the trigger in place with the bait stick, which should be placed so that it is parallel to the ground, with one end touching the trigger (on the side opposite the line coming off the diagonal stick) and the other touching the lower end of the rock.
7. Place food on the bait stick prior to arming the trigger. When a rodent tries to eat the food, it trips the trigger, causing the rock to fall on it.

Box trap

A box trap is ideal for small game and birds. It keeps the animal alive, thus avoiding the problem of having the meat spoil before it's needed for consumption. To construct a box trap, assemble a box from wood and lines, using whatever means are available. Be sure it's big enough to hold the game you intend to catch. Create a two-pin toggle trigger as described for the Twitch-up Strangle Snare, above, by carving L-shaped notches in the center of each stick. For the two-pin toggle to work with this trap, it's necessary to whittle both ends until they're flat. Be sure the sticks you use are long enough to create the height necessary for the animal or bird to get into the box. Take time to make a trigger that fits well.

Set the box at the intended snare site. Secure two sticks at opposite ends on the outside of one of the box's sides. Tie a line to each stick, bring the lines under the box, and secure them to the middle of the lower section of your two-pin toggle. Connect the two-pin toggle together, and use it to raise the side of the box that's opposite the two stakes. Adjust the lines until they're tight and about 1 inch above the ground. Bait the trap. When an animal or bird trips the line, it'll be trapped in the snare.

Apache foot snare

The Apache foot snare is a trap that combines an improvised device that can't easily be removed when penetrated with a simple loop snare made from very strong line. This snare is most often used for deer or similar

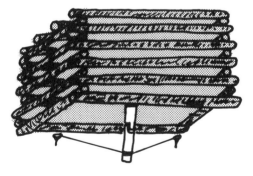

A box trap created using an L-shaped two-pin toggle trigger

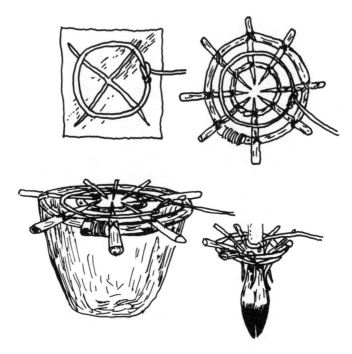

Apache foot snare

animals and is placed on one side of an animal trail obstacle, such as a log. The ideal placement is directly over the depression formed from the animal's front feet as it jumps over the obstacle.

To improvise the device that the animal's foot goes through, gather two saplings, one 20 inches and the other 14 inches long, and eight sturdy branches that are ½ inch in diameter and 10 inches long. Lash each sapling together to form two separate circles, and sharpen one end of each of the eight branches to a blunt point. Place the smaller circle inside the larger, and evenly space the branches over both so that the points approach the center of the inner circle. Lash the sharpened sticks to both of the saplings.

To place the snare, dig a small hole at the depression site, lay the circular device over it, and place the snare line over it. When an animal's foot goes through the device, it will be unable to get it free. As it continues forward, the strong, simple loop snare will tighten down on its foot. When constructing a snare like this, I often use a three-strand braid made from parachute cord, but any strong braid will work. The free end of the snare line should be secured to a large tree or other stable structure. Camouflage the snare with leaves or similar material. Any large animal caught in this snare should be approached with caution.

PREPARING GAME

In order to eat your catch, you'll first need to skin, gut, and butcher most game. Always do this well away from your camp and your food cache. Before skinning an animal, be sure it is dead. Once you're sure, cut the animal's throat, and collect the blood in a container for later use in a stew. If time is not an issue, wait thirty minutes before starting to skin. This allows the body to cool, which makes it easier to skin and also provides enough time for most parasites to leave the animal's hide.

Glove skinning is the method most often used for skinning small game. Hang the animal from its hind legs, and make a circular cut just above the leg joints. Don't cut through the tendon. To avoid dulling your knife by cutting from the fur side, slide a finger between the hide and muscle, and place your knife next to the muscle so that you cut the hide from the inside. Cut

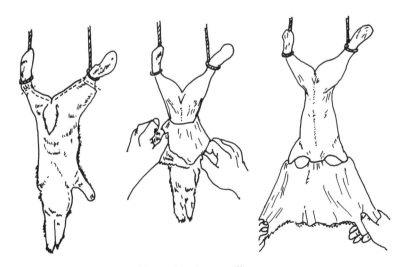

Glove skinning small game

down the inside of each leg, ending close to the genital area, and peel the skin off the legs until you reach the animal's tail. Firmly slide a finger under the hide between the tail and spine until you have a space that allows you to cut the tail free. Follow the same procedure on the front side. At this point, the hide can be pulled down and free from the animal's membrane with little effort. Avoid squeezing the belly, as this may cause urine to spill onto the meat. Pull the front feet through the hide (inside out) by sliding a finger between the elbow and the membrane and pulling the leg up and free from the rest of the hide. Cut off the feet. The head can either be severed or skinned, depending on your talents.

A larger animal can be hung from a tree by its hind legs or skinned while lying on the ground. To hang it by its hind legs, find the tendon that connects the upper and lower leg, and poke a hole between it and the bone. If musk glands are present, remove them. These are usually found at the bend between the upper and lower parts of the hind legs. Free the hide from the animal's genitals by cutting a circular area around them, and then make an incision that runs just under the hide and all the way up to the neck. To avoid cutting the entrails, slide your index and middle fingers between the hide and the thin membrane enclosing the entrails. Use the V between the fingers to guide the cut and push the entrails down and away from the knife. The knife should be held with its backside next to the membrane and the sharp side facing out, so that when used, it cuts the hide from the non-hair side. Next, cut around the joint of each extremity. From there, extend the cut down the inside of each leg until it reaches the midline incision. You should attempt to pull off the hide using the same method as for small game. If you need to use your knife, cut toward the meat so as not to damage the hide. Avoid cutting through the entrails or hide. If skinning on the ground, use the hide to protect the meat, and don't remove it until after you gut and butcher the animal. Once the hide has been removed, it can be tanned and used for clothing, shelter cover, and containers.

To gut an animal, place the carcass, belly up, on a slope or hang it from a tree by its hind legs. Make a small incision just in front of the anus, and insert your index and middle fingers into the cut, spreading them apart to form a V. Slide the knife into the incision between the V formed by your two fingers. Use your fingers to push the internal organs down, away from

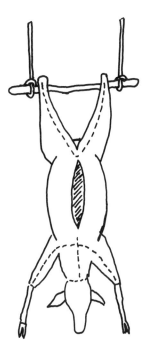

Skinning large game

the knife, and as a guide for the knife as you cut up the abdominal cavity to the breastbone. Avoid cutting the bladder or other internal organs. If they are punctured, wash the meat as soon as possible. Cut around the anus and sex organs so that they will be easily removed with the entrails.

Remove the intact bladder by pinching it off close to the opening and cutting it free. Remove the entrails, pulling them down and away from the carcass. To do this, you will need to sever the intestines at the anus. Save the liver and kidneys for later consumption. If the liver is spotted, a sign of disease, discard all internal organs and thoroughly cook the meat. Cut through the diaphragm and reach inside the chest cavity until you can touch the windpipe. Cut or pull the windpipe free and remove the chest cavity contents. Save the lungs and heart for later consumption. All internal organs can be cooked in any fashion but are best when used in a stew.

If you intend to eat the liver, you'll need to remove the small black sac, the gallbladder, as it's not edible. If it breaks, wash the liver immediately to avoid tainting the meat. Since fat spoils quickly, it should be cut away from the meat and promptly used. The fat is best in soups.

To butcher an animal, cut the legs, back, and breast sections free of one another. When butchering large game, cut it into meal-size roasts and steaks that can be stored for later use. Cut the rest of the meat along the grain into long, thin strips about $1/8$ inch thick, to be preserved by smoking or sun drying. The head meat, tongue, eyes, and brain are all edible, as is the marrow inside bones.

COOKING METHODS

In addition to killing parasites and bacteria, cooking your food can make it more palatable. There are many different ways to prepare game, and some are better than others from a nutritional standpoint. Boiling is best, but only if you drink the broth, which contains many of the nutrients cooked out of the food. Fried food tastes great, but frying is probably the worst way to cook something, as a lot of nutrients are lost during the process.

BOILING

Boiling is the best cooking method. If a container is not available, it may be necessary to improvise one. You might use a rock with a bowl-shaped center, but avoid rocks with a high moisture content, as they may explode. A thick, hollowed-out piece of wood that can be suspended over the fire may also serve as a container. If your container cannot be suspended over the fire, stone boiling is another option. Use a hot bed of coals to heat up numerous stones. Get them really hot. Set your container of food and water close to your bed of hot stones, and add rocks to the water until it begins to boil. To keep the water boiling, cover the top with bark or another improvised lid, and keep it covered except when removing or adding stones. Don't expect a rolling, rapid boil with this process, but a slow, steady bubbling should occur.

BAKING

Baking is the next preferred method of preparing meat to eat. There are several methods you can use to bake game.

Mud baking
When mud baking, there is no need to scale, skin, or pluck a fish or bird in advance, since scales, skin, and feathers will come off the game when the dried mud is removed. Use mud that has a clay texture to it, and tightly seal the fish or bird in it. The tighter the seal, the better it will hold the juices and prevent the meat from drying out. A medium-size bird or trout will usually cook in about fifteen to twenty minutes, depending on the temperature of your coals.

Leaf baking
Wrapping your meat in a nonpoisonous green leaf and placing it on a hot bed of coals will protect, season, and cook the meat. When baking mussels and clams, seaweed is often used; when the shells gap open, they're done. Avoid plants that have a bitter taste.

Underground baking
Underground baking is a good method of cooking larger meals, since the dirt will hold the oven's heat. Dig a hole slightly larger than the meal you intend to cook; it needs to be big enough for your food, the base of rocks, and the covering. Line the bottom and sides with rocks, avoiding rocks with a high moisture content, which may explode, and start a fire over them. To heat rocks that will be used on top of your food, place enough green branches over the hole to support another layer of rocks, leaving a space to add fuel to the fire. Once the green branches burn through and a hot bed of coals is present, remove the fallen rocks. Place green twigs onto the coals, followed by a layer of wetted green grass or nonpoisonous leaves. Add your meat and vegetables, and cover them with more wet grass or leaves, a thin layer of soil, and the extra hot rocks. Then cover the hole with dirt. Small meals will cook in one to two hours, large meals in five to six hours or longer, sometimes days.

Solar oven baking
A solar oven can be improvised from a large cardboard box with a lid, aluminum foil, and clear plastic. Cut 1 inch from the edge of the lid on three of its four sides as shown in the illustration. Make a crease on the uncut side so that the panel can be opened and closed when the lid is closed. Then

Solar oven

False floor of a solar oven

close the lid and open the panel. Cover the inside of the panel and the inside of the box with aluminum foil. Place your food inside a plastic bag or cooking container and into the box. Cover the opening with clear plastic and tape it in place. Position the oven so that it is exposed to the sun, and tilt the open panel into a position that allows maximum reflection into the box. To increase heat absorption and efficiency, consider adding a false floor to the bottom of the box using rocks or similar items.

Frying
Place a flat rock on or next to the fire. Avoid rocks with a high moisture content, as they may explode. Let the rock get hot, and cook on it as you would a frying pan.

Broiling
Broiling is ideal for cooking small game over hot coals. Before cooking the animal, sear its flesh with the flames from the fire. This will help keep the juices, containing vital nutrients, inside the animal. Next, run a non-poisonous skewer—a branch that is small, straight, and strong—along the underside of the animal's backbone. Suspend the animal over the coals, using any means available.

FOOD PRESERVATION

KEEP IT ALIVE
If possible, keep all animals alive until you're ready to consume them. This ensures that the meat stays fresh. A small rodent or rabbit may attract big game, so take measures to protect it from becoming a coyote's meal instead of yours. This doesn't apply, of course, if you're using the rodent as bait.

SUN DRYING
To sun dry meat, you hang long, thin strips in the sun. To keep it out of other animals' reach, run snare wire or line between two trees. If using snare wire, skewer it through the top of each piece of meat before attaching it to the second tree. If using other line, hang it first and then drape the

168 Surviving the Desert

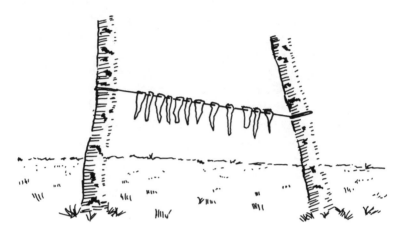

Sun drying meat is an effective method of preserving it for later consumption.

strips of meat over it. For best results, the meat should not touch its other side or another piece.

SMOKING

Smoke long, thin strips of meat in a smoker constructed using the following guidelines:
1. Build a 6-foot-tall tripod from three poles lashed together.
2. Attach snare wire or line around the three poles, in a tiered fashion, so that the lowest point is at least 2 feet above the ground.
3. If using snare wire, skewer it through the top of each slice of meat before extending it around the inside of the next pole. If using other line, hang it first and then drape the strips of meat over it. For best results, the meat should not touch its other side or another piece.
4. Cover the outer aspect of the tripod with any available material, such as a poncho. Avoid contact between the outer covering and the meat. For proper ventilation, leave a small opening at the top of the tripod.
5. Gather an armload of green deciduous wood, such as willow or aspen. Prepare it by either breaking the branches into smaller pieces or cutting the bigger pieces into chips.

6. Build a fire next to the tripod. Once a good bed of coals develops, transfer them to the ground in the center of the smoker. Continue transferring coals as needed.
7. To smoke the meat, place small pieces or chips of green wood on the hot coals. Once the green wood begins to heat up, it should create smoke. Since an actual fire will destroy the smoking process, monitor the wood to ensure that it doesn't flame up. If it does, put it out, but try to avoid disturbing the bed of coals too much. Keep adding chips until the meat is dark and brittle, about twenty-four to forty-eight hours. At this point, it is done.

A smoker is a quick, efficient method of meat preservation.

170 Surviving the Desert

Using a tree cache at night will help protect your food from bears and rodents.

FOOD CACHE

Unless you like to sleep with rodents and perhaps even bigger wildlife, don't store any food in your shelter. During the day, put all food inside a container or cover it. At night, avoid bear and rodent problems by hanging your food in a tree cache. For best results, hang it as high as possible and as far from the trunk as practical.

SURVIVAL TIPS

IF YOU DON'T HAVE WATER, DON'T EAT

The first thing a survivor always seems to worry about is where he or she will get the next meal. Although you may crave food, it will hasten dehydration. Unless you have enough water available, you should not eat.

10
Navigating

Navigation is the ability to get from point A to point B using landmarks and references to identify your position and plan a route of travel. As long as you are able to meet your survival needs, stay put. Rescue attempts are far more successful when searching for a stationary survivor. However, there are three situations when travel from your present location to another might be considered:
- If your present location doesn't have adequate resources to meet your needs, such as for personal protection, sustenance, or signaling.
- If rescue doesn't appear to be imminent.
- If you know your location and have the navigational skills to travel to safety.

MAP AND COMPASS
A map and compass are the basic tools that most backcountry travelers use for navigating in the wilderness. Identifying your location, determining a direction of travel, and avoiding obstacles all require a good understanding of a compass and map nomenclature.

MAP NOMENCLATURE
The particulars of any map's nomenclature can usually be found within its main body and the surrounding margins. For a map to be an effective tool, however, you must become familiar with the one you're using before departing to the wilderness, as there are several different types of maps available. The basic components of most commercial maps are as follows.

Scale
A map's scale refers to the ratio of distances on the map to corresponding distances in real life. The following are commonly used scales:

172 Surviving the Desert

> 1:24,000 scale: every inch on the map represent 24,000 inches in natural terrain.
>
> 1:64,000 scale: every inch on the map represents 64,000 inches in natural terrain.

Series

The series of a map refers to the amount of latitude and longitude displayed on the map. The following are commonly used series:

> 15-minute series: map covers 15 minutes of latitude and 15 minutes of longitude.
>
> 7.5-minute series: map covers 7.5 minutes of latitude and 7.5 minutes of longitude. (It would take four of these maps to cover the same surface area as one 15-minute series map.)

Colors and symbols

The color and symbols on a map denote different things and are very useful in evaluating the terrain. Most common colors and their meanings are as follows:

> Green: woodland.
> White: nonforested areas, such as rocks and meadows.
> Blue: water.
> Black: man-made structures, such as buildings, trails, windmills, and wells.
> Red: prominent man-made items, such as major roads.
> Brown: contour lines.

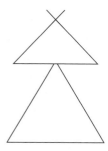 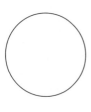

Maps often show man-made wells with one of these two symbols.

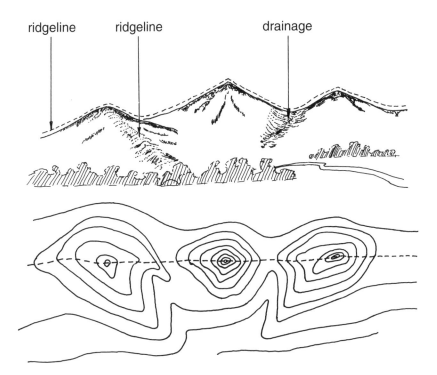

Contour lines are two-dimensional representations of three-dimensional terrain.

Contour lines

Contour lines are imaginary lines, superimposed on a topographic map, that connect points of equal elevation. The contour line interval, usually found in the map's margins, is the distance between two contour lines. The actual distance varies from one map to the next. The following is a basic guide on how to interpret the lay of the land as shown by a map's contour lines:

    Lines close together: steep terrain.
    Lines relatively far apart: gradual elevation gain or loss.
    An area bordered by two steeply sloping sides: canyons.
    Lines form a V pointing toward a higher elevation: drainage.
    Lines form a V pointing away from a higher elevation: ridgeline.

Magnetic variation

An area between two peaks that is usually flat and begins where a V showing a drainage (located between two ridgelines) ends: saddle. Enclosed circle with attached hash marks that point toward the direction of lower ground: depression.

Magnetic variation
The magnetic variation is usually listed at the bottom of a topographic map. An arrow labeled "MN" indicates magnetic north; a second line, with a star at the end, is true north. Maps are set up for true north. This variation, commonly called declination, is valuable in compensating for the difference between true north and magnetic north, which will be your compass heading.

Latitude and longitude lines
Latitude and longitude lines are imaginary lines that encircle the globe, creating a crisscross grid system. These lines help you identify your location.

Latitude lines

Latitude lines are east-west-running lines numbered from 0 to 90 degrees north and south of the equator. The 0-degree latitude line runs around the globe at the equator, and from there the numbers rise to north 90 degrees and south 90 degrees. In other words, the equator is 0 degrees latitude, the North Pole is 90 degrees north latitude, and the South Pole is 90 degrees south latitude. Latitude is noted at the extreme ends of the horizontal map edges.

Longitude lines

Longitude lines are north-south-running lines numbered from 0 to 180 degrees east and west of Greenwich, England, the line commonly referred to as the prime meridian. Longitude lines begin at 0 at Greenwich traveling east and west until they meet at the 180th meridian, which is often referred to as the international dateline. The 0 meridian becomes the 180th

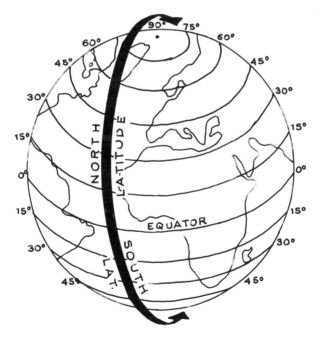

Latitude lines

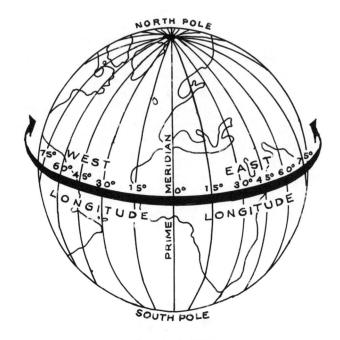

Longitude lines

meridian once it intersects the extreme north and south sections of the globe. Longitude is noted at the extreme ends of the vertical map edges.

Rules for reading latitude and longitude

Both latitude and longitude lines are measured in degrees (°), minutes ('), and seconds ("). There are 60 minutes between each degree and 60 seconds between each minute. It is also important to distinguish north from south when defining your latitude, and east from west for longitude. Whenever giving latitude and longitude coordinates, always read the latitude first.

Latitude: A latitude might read, for example, 45° 30' 30". If north of the equator, your latitude would be 45 degrees, 30 minutes, and 30 seconds north latitude; if south of the equator, 45 degrees, 30 minutes, and 30 seconds south latitude. A latitude line will never be over 90 degrees north or south.

Longitude: A longitude might read, for example, 120° 30' 30". If east of the prime meridian, your longitude would be 120 degrees, 30 minutes,

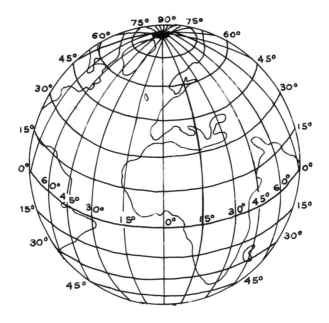

Latitude and longitude intersecting

and 30 seconds east longitude; if west of the prime meridian, 120 degrees, 30 minutes, and 30 seconds west longitude. A longitude line will never be over 180 degrees east or west.

COMPASS NOMENCLATURE
This section describes an orienteering compass and compasses of similar structure, with a circular housing mounted on a rectangular base.

Rectangular base plate
The sides of the base plate have millimeter and inch markings, used to relate a map measurement to that of a relative field distance. The front has a direction-of-travel arrow. The arrow is parallel to the long edge and perpendicular to the short edge. Compass headings are read from the point where the bottom of the direction-of-travel arrow touches the numbers on the edge of the circular compass housing. If the direction-of-travel arrow is not present or centered on the circular housing, your compass will

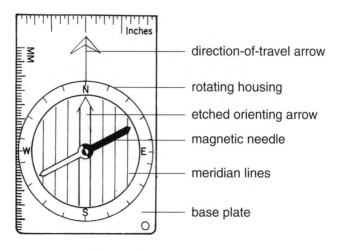

Orienteering compass

probably have a stationary index line, sometimes called an index pointer. This nonmoving short white line is located either on the base plate next to the circular housing or inside the circular housing just beneath the moving numbers (it will be centered on the short wall of the base plate and on the same side of the compass as the direction-of-travel arrow). Headings are read where the numbers touch or pass over this line. The direction-of-travel arrow must always point toward the intended destination when a heading is being taken.

Circular housing
A rotating circular housing sits on the base plate. Its outer ring is marked with the four cardinal points (N, S, E, W) and degree lines starting at north and numbered clockwise to 360 degrees. The bottom of the housing has an etched orienting arrow, which points toward the north marking on the outer ring.

Magnetic needle
The compass needle sits beneath the circular housing. The needle is magnetic, and if you hold the compass close to metal objects, the needle will be

drawn toward them. It floats freely, and one end, usually red, points toward magnetic north (not true north). Magnetic north lies near Prince of Wales Island in northern Canada. Observe in the figure below how the magnetic variance affects readings done in the United States. Notice the line that passes through the Great Lakes and along the coast of Florida. This is an agonic line, which has no variation. In other words, a compass heading of 0 or 360 degrees would point toward both magnetic and true north. The other lines, which are isogonic lines, have variations from true north. The line that extends through Oregon has a variation of 20 degrees east. When this line is extended, the compass bearing of 360 is 20 degrees to the east of true north. The opposite would be true for the line extending through Maine. In this case, a compass bearing of 360 would be 20 degrees west of true north. Because of these variations, adjustments must be made in order to use a map and compass together.

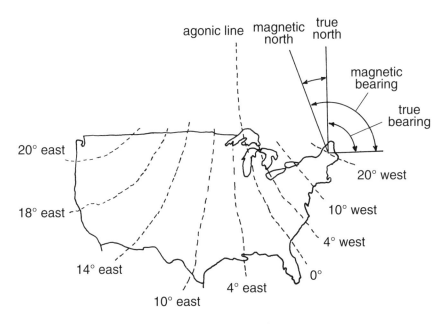

Agonic and isogonic lines

USING YOUR MAP AND COMPASS TOGETHER
Constant awareness and proper use of your map and compass together are the keys to determining your general and specific location.

Determining general location
Anytime you're traveling in the wilderness, you should maintain a constant awareness of your general location, focusing on the surrounding terrain and how it relates to the map you are carrying. If you do this throughout your trip, you shouldn't ever need to use other means of establishing where you are. One way to keep a constant awareness is with dead reckoning. Dead reckoning uses a simple mathematical formula to help you determine your present location:

$$time \times rate = distance$$

Time refers to the amount of time that has passed since you left the last known location, so keep track of your location throughout the day.

The rate of speed you travel is usually measured in miles per hour on land or knots per hour at sea. In a vehicle, keep aware of your speed. The average backpacker travels at a speed of 1 to 3 miles per hour, depending on the weight carried and terrain covered. Take the time to evaluate your speed using known variables. Consider purchasing an electronic pedometer to measure the distance traveled, and use it to determine your average rate of speed by applying it to the following formula, where time and distance are known:

$$distance \div time = rate$$

Once you've determined the distance you've traveled, apply this to your line of travel (heading) from your starting point to figure out your approximate location.

Adjusting your location (latitude and longitude) can be done based on your direction of travel and the distance traveled. There are several other methods you might use to determine your location and direction of travel.

Determining specific location
The first step to determining your specific location is to orient the map. Once this is done, you can either shoot a line of position or triangulate to establish a better idea of your whereabouts.

Orienting your map

Orienting the map aligns its features to those of the surrounding terrain. This process is extremely helpful in determining your specific location.

1. Get to high ground. This will help you evaluate the terrain once the map is oriented.
2. Open the map and place it on a flat, level surface. If possible, protect it from the dirt and moisture with something like a poncho.
3. Rotate the circular housing on the compass until the bottom of the direction-of-travel arrow is touching the true north heading. When doing this, you must account for the area's given declination, the difference between magnetic north (MN) and true north (★). True north is north as represented on a map, and magnetic north is the compass heading. In other words, a 360-degree map heading—true north—is not necessarily a 360-degree compass heading. This variation is usually depicted on the bottom of most topographic maps. If magnetic north is located west of true north, which is the case for most of the eastern United States, you would add your declination to 360 degrees. The resultant bearing would be the compass heading equivalent to true north at that location. If magnetic north is located east of true north, which is the case for most of the western United States, you would

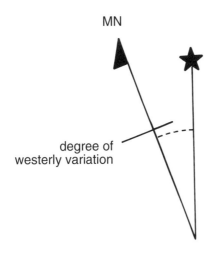

Westerly magnetic variation

182 Surviving the Desert

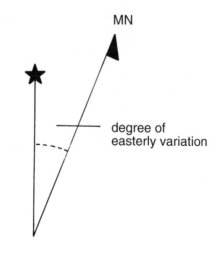

Easterly magnetic variation

subtract your declination from 360 degrees. The resultant bearing would be the compass heading equivalent to true north at that location.
4. Set the compass on the map with the edge of the long side resting next to, and parallel to, the left north-south margins (longitude line). The direction-of-travel arrow should point toward the north end of the map.
5. Holding the compass in place on the map, rotate the map and compass together until the floating magnetic needle is inside the etched orienting arrow of the base plate, with the red portion of the needle forward. This is called boxing the needle.
6. Double-check to ensure that the compass is still set for the variation adjustment, and if correct, weigh down the map edges to keep it in place.
7. At this point, the map is oriented to the lay of the land, and the map features should reflect those of the surrounding terrain.

Line of position to determine your location
A single line of position can be used when at least one prominent land feature can be seen. A prominent land feature includes any easily identified

man-made or natural feature. For best results, get to high ground with 360 degrees of visibility.
1. Orient the map as outlined above.
2. Positively identify the prominent land feature. The following guidelines related to contour, distance, and elevation can help in the identification process.
   - *Contour.* Evaluate the landmark's contour, translating it into a two-dimensional appearance, and search for a matching contour outline on your map.
   - *Distance.* Determine the distance from your present position to the landmark to be identified.

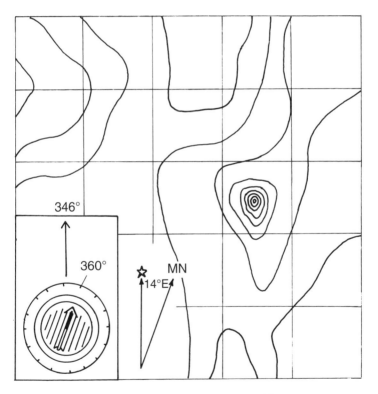

Magnetic variation and orienting the map with a compass

- *In treed terrain.* From 1 to 3 kilometers (approximately 1 to 2 miles), you should be able to see the individual branches of each tree. From 3 to 5 kilometers (approximately 2 to 3 miles), you should be able to see each individual tree. From 5 to 8 kilometers (approximately 3 to 4 miles), the tree will look like a green plush carpet. At greater than 8 kilometers, not only will the trees appear like a green plush carpet, but there will also be a bluish tint to the horizon.
- *In flat, open terrain.* In flat, open terrain, you can calculate a gross distance once a prominent landmark becomes visible on the horizon. To apply this formula, you must first calculate how far you are from the viewed horizon. To do this, take the square root of your eyes' height above the ground and multiply it times 1.23.

$$\sqrt{\text{your eyes' height above ground}} \times 1.23 = \text{distance to the horizon in miles}$$

Next, apply the same formula to the object's height above the ground, and add that to the already calculated horizon distance. The result is a rough estimation of the object's distance from your present location.

$$(\sqrt{\text{height of the object above ground}} \times 1.23) + \text{distance to the horizon in miles} = \text{distance to the object in miles}$$

- *Elevation.* Determine your landmark's height as compared with that of your location.
3. Using your orienteering compass, point the direction-of-travel arrow at the identified landmark, and then turn the compass housing until the etched orienting arrow boxes the magnetic needle (red end forward). At the point where the direction-of-travel arrow intersects the compass housing, read and record the magnetic bearing.
4. Before working further with a topographic map, ensure that it's still oriented.
5. Place the front left tip of the long edge of the compass (or a straight edge) on the identified map landmark, and while keeping the tip in place, rotate the compass until the magnetic needle is boxed (red end

forward). Double-check that your compass heading is correct for the landmark being used.
6. Lightly pencil a line from the landmark down, following the left edge of the compass base plate or straight edge. You may need to extend the line. If you have a protective plastic cover on your map, you can draw on it with grease pencils to avoid exposing the map to moisture and dirt.
7. Your position should be located on or close to the line. For final position determination, evaluate the surrounding terrain and how it relates to your line, along with believed distances to land or light features.

Triangulating to determine your location
Triangulating is a process of identifying your specific location by doing three lines of position. The ideal scenario allows you to positively identify three landmarks that are 120 degrees apart, forming a triangle where the three lines cross. Your position should be located within or around the triangle. For final position determination, evaluate the surrounding terrain and how it relates to the triangle displayed on the map.

ESTABLISHING A FIELD BEARING
Never travel unless you know both your present position and where you intend to go.

Establishing a field bearing with a map and compass
1. Orient your map to the lay of the land.
2. Lightly draw a pencil line from your present location to your intended destination.
3. Place the top left edge of the compass on your intended destination.
4. Rotate the compass until the left edge is directly on and parallel to the line you drew.
5. Rotate the compass housing—keeping the base of the compass stationary—until the floating magnetic needle is boxed inside the orienting arrow (red portion of the needle forward).
6. Read the compass heading at the point where the bottom of the direction-of-travel arrow touches the numbers of the circular compass housing. This heading is the field bearing to your intended destination.

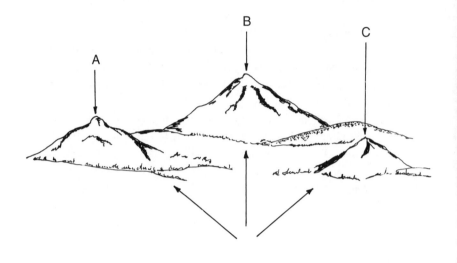
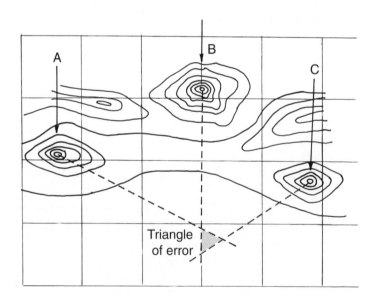

Triangulation of error

Establishing a field bearing with only a compass

1. Holding the compass level, point the direction-of-travel arrow directly at the intended destination site.
2. Holding the compass in place, turn its housing until the magnetic needle is boxed directly over and inside the orienting arrow (red portion of the needle forward).

Using a compass to establish a bearing

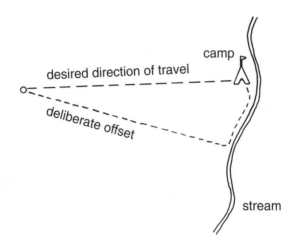

Deliberate offset

3. Read the heading at the point where the bottom of the direction-of-travel arrow touches the numbers on the circular housing. This heading is the field bearing to your intended destination.

Deliberate offset
If your destination is a road, consider a heading with a deliberate offset. In other words, use a field bearing several degrees to one side of your final location. Since it is very difficult to be precise in wilderness travel, this offset will help you decide whether to turn left or right once you intersect the road.

MAINTAINING A FIELD BEARING

Point-to-point navigation
Pick objects in line with your field bearing. Once one point is reached, recheck your bearing and pick another. This method allows you to steer clear of obstacles.

Following the compass
Holding the compass level, while keeping the magnetic needle boxed, walk forward in line with the direction-of-travel arrow.

TRAVEL CHECKLIST

1. *Heading.* Establish the compass heading, or azimuth, to your desired location. Once confident of your azimuth, trust your compass and stay on your heading.
2. *Distance.* Determine the total number of kilometers your route will cover.
3. *Pace count.* Estimate the number of paces it will take to reach your final destination (a pace is measured each time the same foot hits the ground). On fairly level terrain, it takes about 650 paces to go 1 kilometer. On steep terrain, paces will nearly double for each kilometer.
4. *Terrain evaluation.* Evaluate your route's major terrain feature, such as a road or clearing, and determine how many paces it takes to each. By doing this, you'll maintain a constant awareness of your location within your route of travel.
5. *Point description.* Take the time to evaluate the appearance of your final location. This will help you when the time comes to evaluate whether you have had a successful trip.
6. *Estimated arrival time.* Estimating your arrival time will help you set realistic goals on the distance to travel each day.

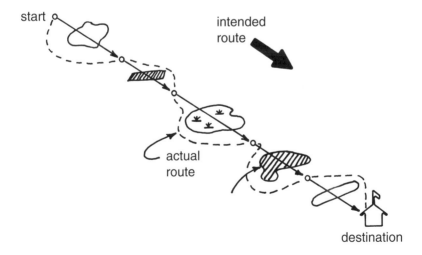

Point-to-point navigation and avoiding obstacles

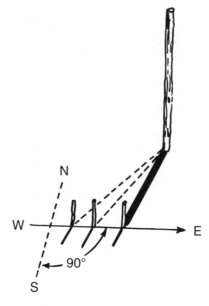

Stick and shadow

DETERMINING DIRECTION USING THE SUN

STICK AND SHADOW
On a flat, level area, clear away all debris from a 3-foot circle until dirt is all that remains. Sharpen both ends of a straight stick, and push one end into the ground until the stick's shadow falls onto the center of the cleared area. Mark the shadow tip with a twig or other material. Wait approximately ten minutes, and place another twig at the shadow tip's new location. Draw a straight line between the two markers, and then another line perpendicular to it. Since the sun rises in the east and sets in the west, the first marking on the shadow line is west and the second one is east.

Generally, in the Northern Hemisphere the sun will be south of your location, and in the Southern Hemisphere the sun will be north of you. This is not always true, however, and depending on where you are, the stick and shadow may not even be an option for use. The following guidelines will help you decide when to use a stick and shadow to determine your cardinal directions:

Navigating

- For the stick and shadow method to be reliable, it cannot be used at greater than 66.6 degrees north or south latitude.
- Between 23.4 and 66.6 degrees north or south latitude, the sun's shadow will be pointing directly north or south (respectively) at local apparent noon.
- Between 0 and 23.4 degrees north or south latitude, the sun can be north or south of your location, depending on the time of year. This poses no problem; simply realize that the first shadow is west and the subsequent shadows move toward the east. A line perpendicular to the east-west line allows you to find which way is north-south.

NAVIGATING WITH A WATCH
Rough cardinal directions can be established using a nondigital watch and the sun.

Northern Hemisphere
Point the watch's hour hand toward the sun. Holding the watch in this position, draw an imaginary line midway between the hour hand and 12:00

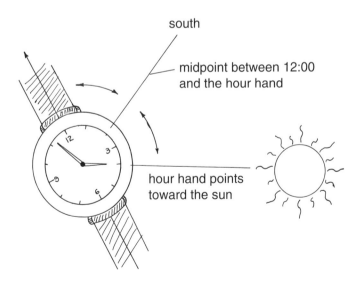

Using a watch in the Northern Hemisphere to determine a southern heading

192   Surviving the Desert

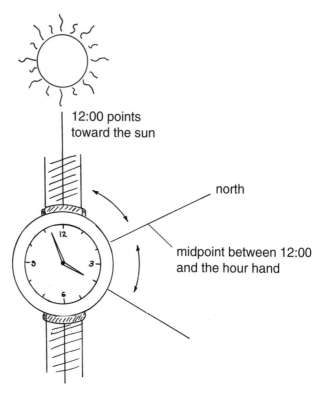

Using a watch in the Southern Hemisphere to determine a northern heading

(1:00 if daylight saving time). This imaginary line represents a southern heading. Draw another line perpendicular to this one to determine the four cardinal directions.

Southern Hemisphere
Point the watch's 12:00 (1:00 if daylight saving time) toward the sun. Holding the watch in this position, draw an imaginary line midway between 12:00 and the hour hand. This imaginary line represents an approximate northern heading. Draw another line perpendicular to the original one to determine the four cardinal directions.

LOCAL APPARENT NOON

At local apparent noon, the sun has reached its highest point and will be due south of you in the Northern Hemisphere and due north of you in the Southern Hemisphere. Local apparent noon is not necessarily the same as a 12:00 reading on your watch—it's unlikely that the sun will always be highest at this time, or even close to it. There are several methods of determining local apparent noon. In order for these methods to work, the horizon height used must be the same for the first recording (sometimes called sunrise) and the second (sometimes called sunset).

To establish local apparent noon, record the exact sunrise and sunset times, based on a twenty-four-hour clock, add them together, and divide the total by 2. Sunrise is when the top of the sun first appears on the horizon; sunset is when the top of the sun disappears on the horizon. This is based on a nonobscured view of the horizon. The resultant figure is your local apparent noon—when the sun is directly north or south, depending on your hemisphere.

(time of sunrise + time of sunset) ÷ 2 = local apparent noon

For example, if sunrise was at 0720 hour and sunset at 1930 hour:

0720 + 1930 = 2650 ÷ 2 = 1325 hour

In this example, 1325 hour, or 1:25 P.M., is local apparent noon, and this figure can be used to help you determine your cardinal directions, since the sun should be directly north or south of you at that time. Given that the sun moves fifteen degrees an hour, you can maintain a course simply by establishing the cardinal directions and then using the sun to adjust your heading throughout the day. If the horizon is obscured by shadows, a kamal device can be used to achieve the same result.

Kamal

A kamal device allows you to create a new horizon above any cloud or haze that obscures your view. Create one by attaching a string to something flat, such as a credit card, and tying a knot in the free end of the string. Place the knot on the string so that the line is tight when the flat plate is held out with an extended arm. To use, place the knot between your teeth, and hold the flat plate out so that its bottom touches the horizon. Your sunrise reading is when the top of the sun first appears at the top of the card; sunset is when the top of the sun disappears below the top of the card. If more height is needed, use a larger card. These figures can be used to determine local apparent noon in the same fashion as described above.

DETERMINING DIRECTION USING THE STARS

NORTHERN HEMISPHERE

In the Northern Hemisphere, Cassiopeia and the Big Dipper are useful tools for helping you find Polaris, the North Star. The Big Dipper looks like a cup with a long handle. Cassiopeia is made up of five stars that form a large W, with its opening facing the Big Dipper. The Big Dipper and Cassiopeia rotate around Polaris, and halfway between these constellations, Polaris can be found. It is located at the very end of the Little Dipper's handle. Contrary to popular belief, it is not the brightest star in the sky, but instead is rather dull. Between 5 and 50 degrees north latitude, Polaris is within 1 degree of true north; between 50 and 60 degrees north, it may be off as much as 2 degrees. Cassiopeia provides the key to this variance.

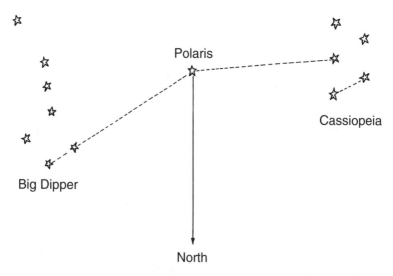

The Big Dipper and Cassiopeia

Navigating   195

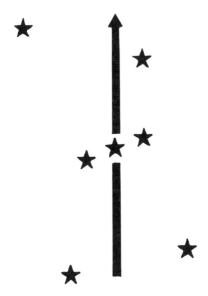

Orion rises in the east and sets in the west.

- Polaris is due north (360 or 0 degree heading) when Cassiopeia is directly above or below its location.
- Polaris is at a 001-degree heading when Cassiopeia is to its right (002 degrees above 50 degrees north latitude).
- Polaris is at a 359-degree heading when Cassiopeia is to its left (358 degrees above 50 degrees north latitude).

When both constellations cannot be seen, you can still find Polaris or determine your cardinal directions with one of the following methods:

- *Big Dipper.* The front end of the Big Dipper's cup is formed by two stars. Extend a line from the uppermost of these two stars approximately four to five times the distance between these two stars to find Polaris.
- *Cassiopeia.* From the center of Cassiopeia, extend a line out approximately four to five times the distance measured between any two of the stars forming the W to find Polaris.
- *Orion's Belt.* Orion the hunter circles the earth directly above the equator. The leading star of Orion's Belt, called Mintaka, rises exactly due east and sets exactly due west. The belt is formed by

three close stars in line at the center of the constellation. When Orion is not directly on the horizon, its east-west path makes it ideal for use with the night stick and shadow technique described below.

SOUTHERN HEMISPHERE

To determine the cardinal directions in the Southern Hemisphere, use the Southern Cross, a constellation with four bright stars that look as though they are the tips of a cross, and the Pointer Stars. The False Cross looks similar to the Southern Cross and may create confusion. The False Cross is less bright than the Southern Cross, and its stars are more widely spaced. In fact, the southern and eastern arms of the actual Southern Cross are two of the brightest stars in the sky. The Pointer Stars are two stars that are side by side and in close proximity to the Southern Cross.

To establish a southern heading, extend an imaginary line from the top through the bottom of the cross. Draw another imaginary line perpendicular to the center of the Pointer Stars. At the point where the lines intersect,

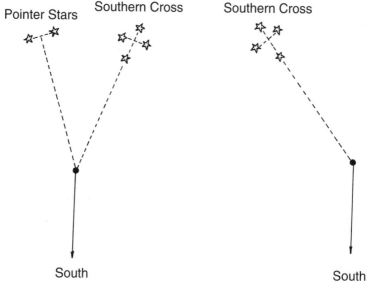

Southern Cross and Pointer Stars

Navigating   197

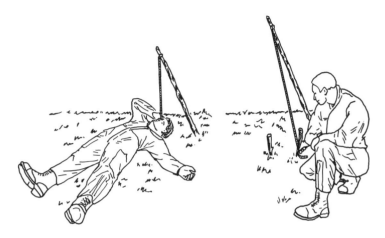

Night stick and shadow

draw a third line straight down toward the ground. This line represents a southern direction.

NIGHT VERSION OF THE STICK AND SHADOW
At night, most travelers use Polaris (the North Star) or the Southern Cross and Pointer Stars to determine the cardinal directions. When these constellations cannot be found, however, you may opt to use stars—located away from the celestial poles—to create cardinal directions. Since these stars generally move from east to west, they can provide the same east-west line as made with the stick and shadow. Find a straight, 5-foot stick and push it into the ground at a slight angle. Tie a piece of line to the top of the stick, ensuring that it is long enough to reach the ground with lots to spare. Lying on your back, position yourself so that you can pull the cord tautly, and hold it next to your temple. Move your body around until the taut line is pointing directly at the selected noncircumpolar star or planet. At this point, the line represents the star's shadow. Place a rock at the place where the line touches the ground, and repeat the process every ten minutes or so. As with the stick and shadow technique, the first mark is west and the second one is east. A perpendicular line will aid you in determining north and south.

GLOBAL POSITIONING SYSTEM (GPS)

A Global Positioning System (GPS) is a tool that can augment solid navigational skills but should *never* replace them. Learn how to use a map and compass before ever laying hands on a GPS. The GPS is an electronic device that works by capturing satellites' signals. To identify your location, given in latitude and longitude coordinates, it must lock on to three satellites; to identify your altitude, it must lock on to four satellites. As with all electronic devices, a GPS is vulnerable to heat, cold, moisture, and sand. Even though satellites' signals are now easier than ever to capture, there are still times when a signal cannot be obtained. In such instances, the GPS is nothing more than added weight in your pack. It's a great tool, but don't rely on it for your sole source of navigation.

SURVIVAL TIPS

DON'T TRY TO NAVIGATE UNLESS YOU KNOW HOW
Navigation is a skill that requires years of practice to become proficient. If you haven't learned to navigate, you have a greater chance of survival if you stay put and wait for rescuers to find you.

TRAVELING AT NIGHT
If you travel at night, use a sturdy, 7-foot-long walking stick. When walking, keep the stick in front of you to protect your face from branches and to feel for irregularities on the ground.

11

Traveling in Hot Climates

Before heading into the desert, leave an itinerary with someone you can trust. Set up check-in times when you will let him or her know you are OK. This insurance is key to a short survival stay versus a long one. If you don't check in, this person can let rescuers know your intended route of travel and initiate a search long before you'd otherwise be missed.

In a desert, limit the amount of travel you do in a survival situation. Leave an area only when it no longer meets your needs, rescue doesn't appear imminent, and you know where you are and have navigational skills. Travel in the desert increases your exposure to aridity and heat, and thus your need for water. To minimize the effects of the desert climate, follow these simple rules:

1. *Travel during the cooler hours.* Rest during the day in a shaded area, and travel in the early morning and late afternoon. Avoid travel in temperatures over 100 degrees F.
2. *Decrease exposure.* Wear proper desert clothing. Cover your body, including your arms and legs, and wear a hat that shades your head and neck. Use sunscreen on exposed areas to prevent sunburn.
3. *Ensure adequate water.* Before departing, drink enough water to ensure that you are well hydrated. So that you stay hydrated during travel in hot desert conditions, carry enough water to drink 1 quart per hour. Also make sure water sources are on your intended route of travel. If not, adjust your course so that they are.

HOW TO CARRY A PACK

The type of pack you carry will depend on personal preference, but in a hot desert climate, an external frame pack is usually preferable because the frame allows air to flow between the pack and your back. To carry a pack on trail, organize your gear so that the heavier items are on top and close to your back. This method places most of the pack's weight on your hips, making it easier to carry. For off-trail travel, organize the pack so that the heavy items are close to the back, from the pack's top to its bottom. With this method, most of the pack's weight is carried by your shoulders and back, allowing you better balance. Pack your larger survival items in the pack so that they can be easily accessed, and carry a smaller survival kit on your person.

BASIC TRAVEL TECHNIQUES

BREAKING TRAIL AND SETTING THE PACE
If traveling in a team, the person breaking trail is working harder than anyone else, and this job needs to be traded off at regular intervals among the members of a team. The leader should set a pace distance and speed that are comfortable for all team members.

KICK-STEPPING
When in scree (small rocks or sand), kick-stepping will make your ascent much easier. Using the weight of your leg, swing the toe of your boot into the rocks, creating a step that supports at least the ball of your foot if going straight up, or at least half of your foot if traversing. When going uphill, lean forward until your body is perpendicular to the earth's natural surface—not that of the hill.

PLUNGE-STEPPING (DOWN-CLIMBING)
Plunge-stepping is similar to kick-stepping, except that you are going downhill and kicking your heels rather than your toes into the slope. Slightly bend the knees, and lean backward until your body is perpendicular to the ground at the base of the hill—not the hill's slope.

TRAVERSING
Traversing, or diagonal climbing, is a quick and easy method for getting up or down a hill. When traversing a hill, it may be necessary to slightly shorten your strides as the grade changes. The same technique can be used to descend a hill.

Traveling in Hot Climates 201

Kick-stepping

Plunge-stepping

202 Surviving the Desert

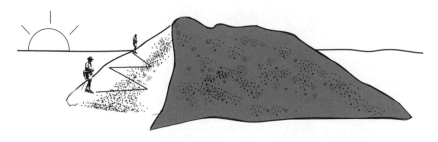

Traversing

REST STEP
When walking uphill, use a rest step, which is done by locking the knee with each step. This process takes the weight off the muscle, allowing it to rest, and places it on the skeletal system. For best results, you'll need to take a short pause with each step.

USING A RIDGELINE TO YOUR ADVANTAGE
When traveling in mountainous terrain, try to stay high on the ridgeline as much as you can. It's better to travel a little farther than to deal with the constant up-and-down travel associated with frequent elevation changes.

TERRAIN ISSUES

LOOSE ROCK SURFACE
Some deserts have a rock floor that is either extremely firm or on occasion soft and brittle. The biggest problem with a firm, rocky desert surface is that it can be hard on your feet and ankles. Constant awareness is necessary to avoid a twist or break that can turn a nice hike into a survival situation.

ROCKY PEAKS
Although it might be tempting to climb a rocky desert peak, be careful. Often people start to climb up a rock only to discover that it is too difficult to reach the top. At this point, most people try to downclimb, and many are surprised to find that the route down is much more difficult than it was

Traveling in Hot Climates 203

going up. If the downclimb becomes too hard, you may slip and get hurt or become stranded on the rock.

CANYONS AND SIMILAR STRUCTURES
A canyon can be like a maze, and unless you have good navigational skills, it's easy to get lost. Before entering a canyon, make sure you can get back out. Do your research. Avoid canyons during peak flash flood seasons. A flash flood in a canyon has the same potential as an avalanche for sweeping you away and taking your life.

CREEKS
If you need to cross a small creek, loosen your pack's shoulder straps and undo your waistband so you can quickly remove the pack if you fall in. Cross the stream in a shallow area by way of a diagonal downstream route. For added stability, use a long walking staff for support. You can also decrease the current's impact on your legs by placing the stick on the upstream side of your position to form a V with you in its center.

LAVA BEDS
Lava beds create unending areas of uneven, hard-to-negotiate terrain. These areas are often filled with obstacles that you cannot go around and that require you to move from one elevated rock to another. Make sure of your footing with each move before you transfer your weight, and always be ready to move on to another location should the one you're on become unstable.

DUNES
Sand dunes form as a result of the wind's movement off the sand. It is far easier to hike on the windward side of a slope, where the sand is packed and more stable. On the leeward side, the sand is soft, and it requires far greater work to get from one point to another.

DRY LAKES
Dry lakebeds are often hard, crusty surfaces that are devoid of visible landmarks. Navigation through these areas should be avoided unless other options are not available. If you do travel through a dry lake, keep an

adequate pace count—it may be the only way you'll be able to identify your current location.

HAZARDS

EXTREME TEMPERATURES
Deserts are known to have temperature extremes, with hot days and cold nights. Travel during midday is not advised; limit your travel to the morning and evening hours.

SANDSTORMS
Sandstorms are a common occurrence in most deserts, and it will be impossible to travel during these times. The potential for eye injury and skin irritation make it important to find or establish a shelter until the storm subsides. In addition, when a storm is present, it doesn't take much for the sand to ruin exposed electronic devices. Take the time to protect them from exposure.

FLASH FLOODS
Avoid dry riverbeds, canyons, and other depressions during the flash flood season. A flash flood has the same life-threatening potential as an avalanche and can easily sweep you up and carry you away. In fact, these floods have been known to originate from up to 100 miles upriver, catching travelers unaware in an area where the sun is shining.

MIRAGES
A mirage is an optical illusion that is often seen in the desert or on a hot road. It occurs when alternate layers of hot and cool air distort light, creating the appearance of a sheet of water where none exists. Mirages make it difficult to identify land features during the day. If mirages are occurring, it may be best to triangulate and plan your routes of travel at dawn or dusk or during moonlight hours.

CAR TRAVEL
You can travel great distances in a vehicle. The opportunity to do this allows you to see a vast array of desert beauty, but it also takes you farther away from civilization. If you break down, you may be so far from town

that your survival odds will greatly depend on what you brought with you and how well you understand the area you are in. Because desert climates are not user-friendly, it's best to leave the driving to someone experienced in such regions. If, however, you decide to do it yourself, leave an itinerary with a friend with specific check-in times and directions on what to do if you don't check in, and observe the following guidelines:

1. Use a vehicle you know has had proper maintenance and is capable of handling the terrain you intend to travel into. Personally check all the fluid levels before you depart, and take along extra just in case.
2. Have the electrical system and battery checked before departing, and carry a spare battery or jump-start battery in addition to jumper cables.
3. Your vehicle's tires should be wide enough to float on sand and have an aggressive tread for adequate traction.
4. Since you have the space, carry at least 5 gallons of water for each person on board, and an additional large survival kit beyond what you might have in your backpack.
5. Before traveling on a rocky desert, make sure your vehicle has enough clearance to handle the terrain.
6. It's best not to drive on a sandy desert, as it's easy to bog down and get stuck. If you must, put your vehicle in four-wheel-drive and let some air out of the tires to increase surface contact. Avoid speed fluctuations and hard turns.
7. Avoid driving on silt and dry lakes unless you have no other option. Silt often has a false floor that won't be able to handle the weight of your vehicle. Dry lakes often have areas with false floors or that are too damp to support your vehicle. In either instance, the vehicle is likely to break through or bog down and get stuck. If you have to cross, take the time to get out of the vehicle and survey the situation. Look for an area where the ground appears solid and is more apt to handle your vehicle's weight.
8. Water crossings should be surveyed before you begin. Make sure the ground is hard enough to support your vehicle and the water is not too deep. In addition, look the selected route over carefully to ensure that your vehicle has enough clearance.
9. If your vehicle breaks down, stay with it. It's the best signal you have and provides a multitude of items that you can use to improvise to meet your various needs.

SURVIVAL TIPS

TRAVEL ONLY IN MORNING AND EVENING
Limit your travel to morning and evening hours. During the day, it's too hot and you increase your odds of developing a heat injury. During the night, your vision becomes impaired and it may be hard to avoid the creatures that are out looking for a tasty meal.

TRAVELING IN SURVIVAL SITUATIONS
If you're in a survival situation and rescue doesn't appear imminent, the area you're in doesn't meet your needs, and you have solid navigational skills, you may elect to travel out to safety. If you do, leave a note for potential rescuers giving your time of departure, route, and intended destination. In addition, mark your trail by tying flags to branches or other landmarks and/or breaking branches.

12
Health Issues

Ultimately, the environment and the amount of time before you return to civilization may have the biggest impact upon any health issues that arise while you're in the desert. The weather may be bad, and the nearest medical facility may be miles from your location. It's highly advisable that you receive adequate first-aid and CPR training, and in no way should you consider this chapter a replacement for that instruction.

GENERAL HEALTH ISSUES

Your ability to fend off an injury or infection plays a significant role in how well you will handle any given survival situation. Proper hydration, nutrition, hygiene, and rest all affect your ability to ward off problems encountered in the wilderness.

STAYING HYDRATED

Without water, you'll die in approximately three to five days. In addition, dehydration will directly affect your ability to make logical decisions about how to handle any given problem. Fluids are lost when the body works to warm itself, when you sweat or engage in intense activity, and when you urinate or defecate. As dehydration starts to set in, you'll begin to have excessive thirst and become irritable, weak, and nauseated. As your symptoms advance, you'll begin having a headache and become dizzy, and eventually your tongue will swell and your vision will be affected. Prevention is the best way to avoid dehydration. This can be accomplished by drinking at least 2 quarts of water during light activity and 4 to 6 quarts (or more on extremely hot days) during more intense activity. If you become dehydrated, decrease your activity, get out of the sun, and drink enough potable water to get your urine output up to at least 1 quart in a twenty-four-hour period.

NUTRITION
Nourishing foods increase morale, provide valuable energy, and replace lost nutrients, such as salt, vitamins, and minerals. Food is not as critical as water, and you may be able to go without eating for several weeks.

HYGIENE
Staying clean not only increases morale, but also helps prevent infection and disease. Methods of staying clean in the wilderness include taking a bath or sun bathing. A sunbath should last from thirty minutes to two hours a day. Keeping your hair trimmed, brushing your teeth and gums, monitoring your feet, and cleaning your cooking utensils after each use are also important tasks that will decrease the risk of infection, illness, or infestation.

REST
Providing the body with proper rest helps ensure that you will have adequate strength to deal with the stress of the initial shock and subsequent trials associated with a survival situation.

TRAUMATIC INJURIES
Traumatic injuries are extremely taxing. Keeping your composure may mean the difference between surviving and not. The treatment of traumatic injuries should therefore follow a logical process. In general, a six-step approach can be taken when evaluating a victim. More details on each treatment follow the outline below.

SIX STEPS FOR A LIFE-THREATENING EMERGENCY
1. *Take charge of the situation.* If in a group, the person with the most medical experience should take charge.
2. *Determine if the scene is safe to enter.* Would entering pose a risk to the subject or rescuer? If so, don't enter until the area is considered safe.
3. *Treat life-threatening injuries.* If the area is unsafe, move the subject to an area that is. Perform a head-to-toe evaluation following this ABCD order:

         A = Airway
         B = Breathing
         C = Circulation and C-spine (neck)
         D = Deadly bleeding
 4. *Treat for shock.* Shock can lead to death, and thus early intervention is a must.
 5. *Secondary evaluation.* Once life-threatening issues have been addressed, perform a secondary survey, evaluating all injuries. Then treat each one.
 6. *Treatment Plan.* Members of the group should discuss how to prevent the subject's condition from worsening and implement an appropriate treatment plan.

AIRWAY, BREATHING, AND CIRCULATION
To successfully treat someone whose airway, breathing, or circulation is compromised, you must know CPR. Learn this prior to departing for the wilderness. CPR can be successful even when the subject has been submerged for up to one hour in cold water.

BLEEDING
There are three types of bleeding. An *arterial bleed* is the most serious of the three and is normally bright red spurting blood. A *venous bleed* can also be very serious and is usually identified as a steady stream of dark red blood. *Capillary bleeds* are minor, and since the vessels are so close to the skin's surface, the dark red blood typically oozes from the site. Basic treatment options are direct pressure, applying pressure to pressure points, and rarely, a tourniquet.

Direct pressure
Do not delay in applying pressure, even if you have to use your hand or finger. If the materials are available for a pressure dressing, pack the wound with several sterile dressings and wrap it with a continuous bandage. The bandage should be snug, but not so snug as to cut off circulation to the rest of the extremity. To ensure that this doesn't occur, regularly check the extremity beyond the wound site for pulses and sensations. If blood soaks through the dressing, apply subsequent dressings directly over the first. If

210 Surviving the Desert

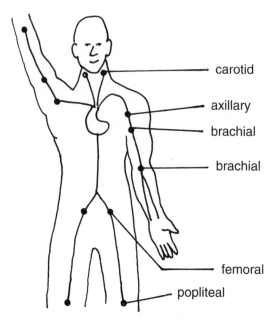

Pressure points

the wound is on an extremity, elevate it above the heart level as long as you can until the bleeding is controlled. In most cases applying direct pressure for ten minutes will stop the bleeding. Leave the dressing in place for two days; thereafter, change it daily.

Pressure points
Applying pressure to a blood vessel between the heart and the wound will decrease the amount of blood loss from the injury site. To be effective, pressure must be applied for about ten minutes. The illustration above shows different pressure points.

Tourniquet
A tourniquet is rarely necessary and should be used only when direct pressure, elevation, and applying pressure to pressure points have all failed or

it's deemed necessary to save a life. The likelihood of losing an extremity from tourniquet use is high; however, once it has been applied, never loosen it. To use a tourniquet, apply a 3- to 4-inch band 2 inches above the wound, so that it is between the wound and the heart. After wrapping the band around the limb several times, tie it into a square knot with a sturdy stick placed in the knot's center, tighten the tourniquet by turning the stick until the blood flow comes to a stop, and secure the stick in place. Mark the victim's head with a big T and note the time when the tourniquet was applied.

SHOCK
Shock is a direct result of the body's inability to provide a sufficient blood supply to the vital organs. If not treated, it could ultimately lead to death. Symptoms of shock include pale, cold, clammy skin; a weak, rapid pulse; and feelings of restlessness, disorientation, and faintness. All injuries, no matter how small, can potentially lead to shock, and all victims should be treated as if shock were present. To treat, control the patient's heat loss by covering him or her with any form of dry insulating material and provide insulation from the ground. If hypothermia is present, treat it. If conscious, lay the victim on his or her back. If unconscious, lay him or her on the side, in case of vomiting. Elevate lower extremities 8 to 12 inches, except when there is a serious head, neck, chest, or abdominal injury. For conscious victims with head or chest injury, raise the upper torso about 15 degrees toward a sitting position.

HEAD INJURIES
Signs and symptoms of a head injury include bleeding, increasing headache, drowsiness, nausea, vomiting, unequally dilated pupils, and unconsciousness. To treat, immobilize the neck if a neck injury is suspected; monitor for any change in mental status; and treat for shock as above.

SPINAL INJURIES
Symptoms of a spinal injury include pain, numbness, tingling, decreased sensation or lack of feeling in extremities, and the inability to move the body below the injury site. Immobilize the neck and body if spinal injury is suspected (any firm, flat surface will do) and treat for shock as above.

ABDOMINAL INJURIES
Signs and symptoms of an abdominal injury include bleeding, abdominal wall bruising, pain, drowsiness, nausea, and vomiting. To treat an open intestinal wound, rinse away any dirt and debris with a mixture of 1 quart sterile water and 1 teaspoon salt. After cleaning the area, cover it with a clean dressing moistened with the above solution. It's extremely important to prevent the intestines from drying out. Victims of both open and closed abdominal injuries should be treated for shock as above.

CHEST INJURIES
Symptoms of a chest injury can vary tremendously, depending on the cause or problem. As a general rule, victims may have pain, cough, be short of breath, have an irregular breathing pattern (rapid or slow), exhibit anxiety, or have cyanosis (bluing around lips and fingers). An open chest wound should be covered with a piece of plastic or other airtight material; dressing material may be used but it isn't as effective. Tape the covering on three sides, allowing air to escape but not enter the opening. If the victim's breathing pattern worsens, remove the patch. Victims of both open and closed chest injuries should be treated for shock as above.

FRACTURES

Closed fractures
Signs and symptoms of a closed fracture include site deformity, swelling, pain, and an inability to bear weight on the affected extremity. To treat, clean all open wounds, and apply a splint that immobilizes the extremity one joint above and below the fracture site. A splint can be improvised by using strong branches that are held in place with 1-inch-wide bands of clothing or similar material. Once you've applied a splint, monitor for any changes in circulation or sensation. When in doubt about whether something is broken or not, treat it as if it is.

Open fracture
An open fracture has all the signs and symptoms of a closed fracture, with the addition of bone protruding through the skin. Don't push bone ends back in or handle them during the treatment process. To treat, rinse away

any dirt and debris with a mixture of 1 quart sterile water and 1 teaspoon salt. After cleaning the bone and the surrounding area, cover the bone ends with a clean dressing moistened with the above solution. It's extremely important to prevent the bone ends from drying out. Secure the dressing in place, splint the fracture, and monitor the extremity for any changes in circulation or sensation.

SPRAINS AND STRAINS
With a sprain, the area of pain is over a joint. With a strain, the area of pain is over a muscle. A sprain will have symptoms similar to those of a closed fracture and should be treated as if a fracture exists. A strain results in localized muscle tenderness, usually as a result of overuse or trauma. To treat a strain, apply moist heat and discontinue any activity that seems to make it worse.

BURNS
Burns are rated by depth as first, second, or third degree, each indicating increasingly deeper penetration. A first-degree burn causes superficial tissue damage, sparing the underlying skin, and is similar in appearance to a sunburn. A second-degree burn causes damage into the upper portion of the skin, with resultant blister formation that is surrounded by first-degree burn damage. A third-degree burn causes complete destruction of the skin's full thickness and often beyond. In addition, first- and second-degree burns are usually present.

   To treat burns, cool the skin as rapidly as possible and for at least forty-five minutes. This is extremely important, since many burns continue to cause damage for up to forty-five minutes, even after the heat source has been removed. Remove clothing and jewelry as soon as possible, but don't remove any clothing that is stuck in the burn. Never cover the burn with grease or fats, as they will only increase the risk of infection and are of no value in the treatment process. Clean the burn with water (preferably sterile), apply antibiotic ointment, and cover it with a clean, loose dressing. To avoid infections, leave the bandage in place for six to eight days. After that time, change the bandage as necessary. If the victim is conscious, fluids are a must. Major burns cause a significant amount of fluid loss, and ultimately the victim will go into shock unless these fluids are replaced. If pain medications are available, use them. Burns are extremely painful.

FOREIGN BODIES IN THE EYE

Most eye injuries encountered in the wilderness are a result of dust, dirt, or sand blown into the eye by the wind. Symptoms include a red and irritated eye, light sensitivity, and pain in the affected eye. To treat, first look for any foreign bodies that might be causing the irritation. The most common site where dirt or dust can be found is just under the upper eyelid. Invert the lid and try to find and remove the irritant. If you're unable to isolate the cause, rinse the affected eye with clean water for at least ten to fifteen minutes. When rinsing, keep the injured eye lower than the uninjured to avoid contaminating the other eye. Apply ophthalmic antibiotic ointment, if available, to the affected eye.

WOUNDS, LACERATIONS, AND INFECTIONS

Clean all wounds, lacerations, and infections, and apply antibiotic ointment, a dressing, and a bandage daily.

BLISTERS

Blisters result from the constant rubbing of your skin against a sock or boot. The best treatment is prevention. Monitor your feet for hot spots or areas that become red and inflamed. If you develop a hot spot, apply a wide band of adhesive tape across and well beyond the affected area. If you have tincture of benzoin, use it. It will make the tape adhere better, and it also helps toughen the skin. To treat a blister, cut a blister-size hole in the center of a piece of moleskin, and place it so that the hole is directly over the blister. This will take the pressure off the blister and place it on the surrounding moleskin. Avoid popping the blister. If it does break open, treat it as an open wound, applying antibiotic ointment and a bandage.

THORNS, SPLINTERS, AND SPINES

Thorns and splinters are often easy to remove. Cactus spines, however, hook into the skin, and in most cases you'll need a pair of tweezers or pliers to get them out. If you can't pull out the spines, don't panic. They often come out on their own over a period of several days. Whether or not you remove them, prevent infection and protect the area by applying antibiotic ointment, a dressing, and a bandage.

FISHHOOK INJURIES
A fishhook can be left in place for short periods of time if you know rescue is coming. If rescue is not expected within several hours, however, it should be removed. The easiest way to do this is to advance the hook forward until the barb clears the skin, cut off the barb, and reverse the hook back out.

HEAT RASH
Heat rashes often occur in moist, covered areas of the body. These bumpy red irritants can be pretty uncomfortable. To treat, keep the area clean and dry, and air it out as much as you can. If you have hydrocortisone 1 percent cream, apply a thin layer to the rash twice a day.

ENVIRONMENTAL INJURIES AND ILLNESSES
The environment challenges us in many different ways, and it needs to be respected. Realize that it cannot be conquered. Adapting and being properly prepared will play a significant role in surviving nature's sometimes awesome power.

HEAT INJURIES

Sunburn
Prevent sunburn by using a strong sunscreen before exposure to the hot sun. If sunburn should occur, apply cool compresses, avoid further exposure, and cover any areas that have or may become burned.

Muscle cramps
Muscle cramps are a result of excessive salt loss from the body, exposure to a hot climate, or excessive sweating. Painful muscle cramps usually occur in the calf or abdomen while the victim's body temperature is normal. To treat, immediately stretch the affected muscle. The best way to prevent recurrence is to consume 2 to 3 quarts of water per day when engaged in

minimal activity, and 4 to 6 quarts per day when in extreme cold or hot environments or perhaps even more during heavy activity.

Heat exhaustion
Heat exhaustion is a result of physical activity in a hot environment and is usually accompanied by some component of dehydration. Symptoms include feeling faint or weak, cold and clammy skin, headache, nausea, and confusion. To treat, rest in a cool, shady area and drink plenty of water. Since heat exhaustion is a form of shock, you should lie down and elevate your feet 8 to 12 inches.

Heatstroke
Heatstroke occurs when the body is unable to adequately lose its heat. As a result, body temperature rises to such high levels that damage to the brain and vital organs occur. Symptoms include flushed dry skin, headache, weakness, lightheadedness, rapid full pulse, confusion, and in severe cases, unconsciousness and convulsions. Heatstroke is a true emergency and should be avoided at all costs. Immediate treatment is imperative. Immediately cool the victim by removing his or her clothing and covering the body with wet towels or by submersion in water that is cool but not icy. Fanning is also helpful. Be careful to avoid cooling to the point of hypothermia.

Hyponatremia
Hyponatremia is a potentially fatal condition that can occur under extremely hot conditions. It is caused by a lack of sodium in the blood and frequently occurs when someone drinks too much water while losing high levels of body salt through sweating. Symptoms are dizziness, confusion, cramps, nausea, vomiting, fatigue, frequent urination, and in extreme conditions, coma and even death. To treat, stop all activity, move to a shaded area, treat for shock, and have the victim eat salty foods along with small quantities of lightly salted water or sports drinks. If the victim's mental alertness decreases, seek immediate help.

COLD INJURIES
Because deserts often have extremely cold nights, cold injuries also may be a threat.

Hypothermia

Hypothermia refers to an abnormally low body temperature. Symptoms include uncontrollable shivering, slurred speech, abnormal behavior, fatigue and drowsiness, decreased hand and body coordination, and weakened respiration and pulse. Body heat may have been lost through radiation, conduction, evaporation, convection, and/or respiration. The best treatment is prevention through avoidance of exposure and early recognition. Dressing appropriately for the environment and maintaining adequate hydration can help you avoid most problems with hypothermia. If hypothermia does occur, it should be treated without delay. Begin treatment by stopping continued heat loss. Get out of the wind and moisture, and put on dry clothes, a hat, and gloves. In you have a sleeping bag, take off your clothes, fluff the bag, and climb inside. If an extreme case, someone else should disrobe and climb inside the bag with you. If conscious, you should consume warm fluids and carbohydrates.

Frostbite

Frostbite commonly affects toes, fingers, and the face. The best treatment is prevention. Following the guidelines of the COLDER acronym in chapter 4 and understanding how heat is lost can help ensure that frostbite doesn't occur. Frostbite can be superficial or deep. *Superficial frostbite* causes cold, numb, and painful extremities that appear white or grayish in color. To treat, rewarm the affected part with your own or someone else's body heat: place hands in the armpits, feet on another person's abdomen. Cover other exposed areas with loose, layered material. Never blow on your hands; the resultant moisture will cause the skin to freeze or refreeze. *Deep frostbite* causes your skin to take on a white appearance, lose feeling, and become extremely hard. Should you sustain a deep frostbite injury, don't attempt to rewarm it. Not only would this cause extreme pain, but whereas you can walk on a frostbitten limb, a rewarmed limb would be rendered useless. Prevent any further freezing and injury from occurring by wearing proper clothing and avoiding further exposure to the elements.

IMMERSION INJURIES

Immersion foot is a direct result of long-term exposure of the feet to wet socks. It usually takes several days to weeks of this exposure before damage occurs. Symptoms include painful, swollen feet or hands that have a waterlogged appearance. Since immersion injuries can be debilitating, it's best to prevent them by changing wet socks quickly, not wearing tight clothing, and increasing foot circulation with regular massages. If an immersion injury develops, treat it by keeping the feet dry and elevated. Since rubbing may result in further tissue damage, pat wet areas dry.

SUN BLINDNESS

Sun blindness is a result of exposure of the eyes to the sun's ultraviolet rays. It most often occurs in areas where sunlight is reflected off sand, snow, water, or light-colored rocks. The resultant burn to the eyes' surface can be quite debilitating. Symptoms include bloodshot and tearing eyes, a painful and gritty sensation in the eyes, light sensitivity, and headaches. Prevention by wearing 100 percent UV sunglasses is a must. If sun blindness does occur, avoid further exposure, apply a cool wet compress to the eyes, and treat the pain with aspirin as needed. If symptoms are severe, apply an eye patch for twenty-four to forty-eight hours.

ALTITUDE ILLNESSES

As your elevation increases, so does your risk of developing a form of altitude illness. As a general rule, most mountaineers use the following three levels of altitude to determine their risk of medical problems: high altitude, 8,000 to 14,000 feet; very high altitude, 14,000 to 18,000 feet; extreme high altitude, 18,000 feet and above. Since most travelers seldom venture to heights greater than 14,000 feet, most altitude illnesses are seen in the high-altitude range.

As your altitude increases, your body goes through compensatory changes, including increased respiratory and heart rates, increased red blood cell and capillary production, and changes in the body's oxygen delivery capacity. Most of these changes occur within several days to weeks of exposure at high altitudes. To diminish the effects of altitude, make a gradual ascent, avoid heavy exertion for several days after rapidly ascending to high altitudes, and ingest only small amounts of salt. If you have a history of pulmonary edema or worse, consider taking Diamox (acetazolamide), a prescription medication that is contraindicated for individuals with kidney, eye, or liver disease. The usual dose is 250 milligrams taken two to four times a day. It's started twenty-four to forty-eight hours prior to ascent, and continued while at high altitude for forty-eight hours or as long as needed.

High-altitude illnesses are a direct result of a reduction in the body's oxygen supply. This reduction occurs in response to the decreased atmospheric pressure associated with higher elevations. The three illnesses of high altitude are acute mountain sickness, high-altitude pulmonary edema, and high-altitude cerebral edema.

Acute mountain sickness
Acute mountain sickness usually occurs as a result of decreased oxygen supply to the brain at altitudes greater than 8,000 feet. Symptoms include headache, fatigue, dizziness, shortness of breath, decreased appetite, nausea and vomiting, feelings of uneasiness, cyanosis (bluing around lips and fingers), and fluid retention in face and hands. In severe cases, there may be evidence of some impaired mental function, such as forgetfulness, loss of memory, decreased coordination, hallucinations, or even psychotic behavior. To prevent acute mountain sickness, allow time to acclimatize by keeping activity to a minimum for the first two to three days after arriving at elevations greater than 8,000 feet; avoid alcohol and tobacco; eat small, high-carbohydrate meals; and drink plenty of fluids. If symptoms are severe and oxygen is available, give 2 liters per minute through a face mask for a minimum of fifteen minutes. If symptoms persist or worsen, descend at least 2,000 to 3,000 feet; this is usually enough to relieve symptoms.

High-altitude pulmonary edema (HAPE)
High-altitude pulmonary illness is an extremely common and dangerous type of altitude illness that results from abnormal accumulation of fluid in the lungs. It most often occurs when a climber rapidly ascends above 8,000 feet and, instead of resting for several days, immediately begins performing strenuous activities. Symptoms include signs of acute mountain sickness, shortness of breath with exertion that may progress to shortness of breath at rest as time goes by, shortness of breath when lying down (this symptoms usually makes it difficult for the victim to sleep), and a dry cough that, in time, will progress to a wet, productive, and persistent cough. If symptoms progress, the climber may show symptoms of impaired mental function similar to those seen in acute mountain sickness. If the climber

becomes unconscious, death will occur within several hours unless a quick descent is made and oxygen treatment started. An early diagnosis is the key to successfully treating pulmonary edema. Once identified, immediately descend a minimum of 2,000 to 3,000 feet, or until symptoms begin to improve. Once down, rest for two to three days and allow the fluid that has accumulated in the lungs to be reabsorbed by the body. If oxygen is available, administer it, via a tight-fitting face mask, at 4 to 6 liters per minute for fifteen minutes, and then decrease its flow rate to 2 liters per minute. Continue using the oxygen for an additional twelve hours if possible. If the victim has moderate to severe HAPE, he or she should be evacuated to the nearest hospital as soon as possible. If prone to HAPE, it may be worth trying Diamox prior to the climb (see above).

High altitude cerebral edema (HACE)
High-altitude cerebral edema is swelling of the brain, and it most often occurs at altitudes greater than 12,000 feet. Edema forms as a consequence of the body's decreased supply of oxygen, a condition known as hypoxia. Symptoms include signs of acute mountain sickness, headache that is usually severe and unrelenting, abnormal mental function (confusion, loss of memory, poor judgment, or hallucinations), and ataxia (poor coordination). Severe cases can result in coma or even death. Early recognition is of the utmost importance in saving someone who develops HACE. A person with a severe chronic headache with confusion and/or ataxia must be treated for high-altitude cerebral edema—a true emergency. To treat, descend immediately. If the victim is ataxic or confused, he or she will need help. If oxygen is available, administer it as described above for a HAPE victim. Even if the victim recovers, he or she shouldn't return to the climb. If a person becomes unconscious or has severe symptoms, all efforts should be made for an air evacuation to the nearest hospital.

BOWEL DISTURBANCES
Bowel disturbances in the wilderness are common and include diarrhea and constipation.

Diarrhea
Diarrhea is a very common occurrence in a survival situation. In the desert, diarrhea can lead to dehydration and hyponatremia. Some common

causes are changes in water and food consumption, drinking contaminated water, eating spoiled food, eating off dirty dishes, and fatigue or stress. Diarrhea is almost always self-limiting, and unless you have antidiarrhea medications, treatment should consist of supportive care. Consume clear liquids for twenty-four hours, and follow with another twenty-four hours of clear liquids plus bland foods.

Constipation
Constipation is common in a survival setting. To treat, drink fluids and exercise. Laxatives are contraindicated and rarely needed.

SURVIVAL STRESS
The effects of stress in a survival situation cannot be understated. To decrease its magnitude, you must not only understand it, but also prevail over it. The environment, your condition, and the availability of materials will either raise or decrease the amount of stress you'll experience. The most important key to overcoming survival stresses is the survivor's will. The will or drive to survive is not something that can be taught. However, your will is directly affected by the amount of stress associated with a survival situation. Observing the six Ps of survival—proper prior preparation prevents poor performance—will help alleviate some of this stress. In addition, following my three-step approach to survival—keeping a clear head and thinking logically, prioritizing your needs, and improvising—will help raise your comfort level.

SURVIVAL TIPS

FAMILIARIZE YOURSELF WITH THE AREA
Before planning a weeklong trip into the desert, become familiar with the area by taking short trips that pose little risk. Become accustomed to its hazards and the climate's effects on you. Also take the time to talk with the local authorities about water sources, terrain hazards, flood potential, and dangerous creatures and humans.

13
Desert Creatures

SNAKE, LIZARD, AND ANIMAL BITES

SNAKEBITES
Treat all snakebites as though poisonous unless you can positively identify the snake as nonpoisonous. Of those that are poisonous, few are ever fatal or debilitating with proper medical intervention. Poisonous snakebites are often categorized as hemotoxic, damaging blood vessels and causing hemorrhage, or neurotoxic, paralyzing nerve centers that control respiration and heart action. Common signs that envenomization has occurred include some of the following:

Hemotoxic envenomization (rattlesnake, puff adder, sidewinder, sand viper, horned viper)

Immediate: one or more fang marks and bite site burning.

5 to 10 minutes: mild to severe swelling at the bite site.

30 to 60 minutes: numbness and tingling of the lips, face, fingers, toes, and scalp. If these symptoms occur immediately following a bite, they are likely due to anxiety and hyperventilation.

30 to 90 minutes: twitching of the mouth, face, neck, eye, and bitten extremity. In addition, the victim may develop a metallic or rubbery taste in the mouth.

1 to 2 hours: sweating, weakness, nausea, vomiting, chest tightness, rapid breathing, increased heart rate, palpitations, headache, chills, confusion, and fainting.

2 to 3 hours: the area begins to appear bruised.

6 to 10 hours: large blood blisters often develop.

6 to 12 hours: difficulty breathing, increased internal bleeding, and collapse.

Neurotoxic envenomization (coral snake, cobra, kraits, mambas)
    Immediate: bite site burning may or may not occur, and only a small amount of localized bruising and swelling is often noted.
    Within 90 minutes: numbness and weakness of the bitten extremity.
    1 to 3 hours: twitching, nervousness, drowsiness, giddiness, increased salivation, and drooling.
    5 to 10 hours: slurred speech, double vision, difficulty talking and swallowing, and impaired breathing.
    10 hours or more: death is often the end result without medical intervention.
    Snakebite treatment centers on getting the victim to a medical facility as fast as you safely can. In doing so, follow these basic treatment guidelines to increase survivability:

- Have the victim move out of the snake's range, then stop, lie down, and stay still. Physical activity will increase the spread of the venom. If you can do so safely, try to identify the kind of snake. If you can kill the snake, do so and bring it along for identification purposes. Protect yourself from accidental poisoning by cutting off the head and burying it. Details on how to kill a snake are given in chapter 9.
- Remove the toxin from the wound site as soon as possible using a mechanical suction device, following the manufacturer's instructions, or by squeezing for thirty minutes. Don't cut and suck. This will hasten the spread of the poison and also expose the small blood vessels under the aid giver's tongue to the venom.
- Remove all jewelry and restrictive clothing from the victim.
- Clean the wound, and apply a dressing and bandage. Do not pour alcoholic beverages on the wound, and do not apply ice. Circulation to the site is already impaired, and applying ice may cause symptoms similar to severe frostbite. If the bite is on an extremity and you are more than two hours from a medical facility, use a pressure dressing over the wound or constrictive band—not a tourniquet—placed 2 inches above the site, between it and the heart. This will help restrict the spread of the poison.
    —*Pressure dressing.* Place a clean dressing over the bite and cover it with an elastic wrap that encircles the extremity. The wrap should be about 10 inches wide and centered firmly on

top of the bite site. Although it should be snug, make sure it isn't so tight that it cuts off the circulation to the fingers or toes. Nail bed capillary refill should return with two to three seconds, and the victim should have normal feeling beyond the dressing site. To assess capillary refill, press on the victim's fingernail and count how many seconds it takes for it to resume its pink color once released.

— *Constrictive band.* A constrictive band is not a tourniquet. It is used to slow down the flow in the superficial veins and lymph system. Use any material that allows you to create a 4-inch-wide band, wrapping it around the extremity so that it is between the bite and the heart. If limb swelling makes the band too tight, it can be moved up the extremity.

- After a dressing or band is applied, splint the extremity. The victim should keep the wound site positioned below the level of the heart.
- Have the victim drink small amounts of water.
- Transport the victim to the nearest hospital.

It's best to take precautions to avoid snakebites in the first place. Avoid known habitats like rocky ledges and woodpiles. If you see a snake, leave it alone unless you intend to kill it for food. Carry a walking stick that can be used for protection, and wear boots and full-length pants.

LIZARD BITES

The Gila monster and its cousin, the Mexican beaded lizard, are the only two known species of venomous lizards. Both are similar in appearance and habits, but the Mexican beaded lizard is slightly larger and darker. The Gila monster averages 18 inches in length and has a large head, stout body, short legs, strong claws, and a thick tail that acts as a food reservoir. Its skin is coarse and beadlike, with a marbled coloring that combines brown or black with orange, pink, yellow, or dull white. Most of the lizard's teeth have two grooves that guide the venom, a neurotoxin that affects the nervous system, from glands in the lower jaw. The venom enters the wound as the lizard chews on its victim. Although a bite can be fatal to humans, it usually isn't. Treatment for lizard bites is the same as that for a snakebite.

ANIMAL BITES
Thoroughly clean the site and treat it as any other open wound.

INSECTS, CENTIPEDES, SPIDERS, AND SCORPIONS
Clean any sting or bite that cannot be identified, and use antihistamines when appropriate. Remove any stingers using whatever means are suitable. In most cases, this is done by scraping a knife or similar item 90 degrees across the stinger. Monitor for secondary infection and treat with antibiotics if one occurs.

BEE OR WASP
If stung, immediately remove the stinger by scraping the skin, at a 90-degree angle, with a knife or your fingernail. This will decrease the amount of venom that is absorbed into the skin. Applying cold compresses and/or a cool paste made of mud or ashes will help relieve the itching and pain. To avoid infection, don't scratch the stinger site. Carry along a bee sting kit, and review the procedures of its use prior to departing for the wilderness. If someone has an allergic anaphylactic reaction, it's necessary to act fast. Using the medications in the bee sting kit and following basic first-aid principles will reverse the symptoms associated with this type of reaction in most cases. Regardless of results, it's best to get the victim to the nearest hospital as soon as possible when anaphylaxis occurs.

KISSING BUG (CONENOSE BUG)
The kissing bug is dark brown to black with reddish-orange spots on the abdomen and measures ½ to 1 inch long. It has a cone-shaped head on a long body and three pairs of legs. It usually bites and feeds on the blood of its victim when the victim is asleep. The name kissing bug is derived from the fact that the bug often bites its victims on the lips. These bugs often live inside rodent and birds' nests and are seen in spring and early summer. The bites may be painful and cause redness, swelling, and itching. In some instances, sensitive individuals can have a serious allergic reaction that causes severe itching, rash, nausea, vomiting, and breathing problems. Anaphylactic reaction can occur in very sensitive people. Treatment involves cleaning the site and using antihistamines when necessary. If allergic, use a bee sting kit and seek immediate medical attention.

ANTS

Ants, especially fire ants, can produce very painful bites that often leave small, clear blisters on the skin. The biggest concern aside from pain is avoidance of secondary infection. Clean the bite with soap and water, and use antihistamines if needed. If an infection occurs, treat as any other infection. If allergic, use a bee sting kit and seek immediate medical attention.

TICKS

Remove a tick by grasping it at the base of its body, where its mouth is attached to the skin, and applying gentle backward pressure until it releases its hold. If its head isn't removed, apply antibiotic ointment, bandage, and treat as any other open wound.

MOSQUITOES AND FLIES

To minimize the number of bites you'll experience from these pesky insects, use insect repellent and cover the body's exposed parts with clothing or mud. Insects that carry parasitic, viral, and bacterial agents transmit vector-borne diseases. Common diseases are malaria (in the tropics) and West Nile virus. Since wild and domestic birds carry West Nile virus, it appears to have no boundaries; mosquitoes become carriers when they bite an infected bird. The risk of getting the virus is seasonal, beginning in the spring and reaching its peak in mid-to late August. Approximately 80 percent of those infected will not have any symptoms. When symptoms do occur, they usually only last a few days and include fever, headache, muscle aches, backache, skin rash, and swollen lymph glands. In rare cases the infection can lead to an infection in the brain or its lining. Treatment is supportive and includes rest, fluids, and pain control. If you think you have been infected, you should seek out medical care as soon as possible.

CENTIPEDES AND MILLIPEDES

Centipedes inject venom using fanglike front legs. Millipedes have toxins on their bodies that, when touched, are highly irritating. Both can cause redness, swelling, and pain to the bite site. If bitten, clean the area with soap and water. Use a pain medication if needed.

SPIDERS

Desert spiders avoid the searing heat by taking cover in burrows or under rocks, emerging at night to eat. The most prominent dangerous spiders of the deserts are the black widow, brown recluse, and tarantulas.

Black widow

The black widow's venom is fifteen times as toxic as the venom of the prairie rattlesnake, and it is considered the most venomous spider in North America. However, black widow spiders inject only a relatively small amount of venom and are not usually deadly to adults. Only the female spider is venomous. The black widow female is shiny black and often has a reddish hourglass shape on the underside of her spherical abdomen. Her body is about 1½ inches long. The adult male is harmless, about half the female's size, with a smaller body and longer legs, and usually has yellow and red bands and spots over the back. The black widow's bite may be painless and go unnoticed. Symptoms may include muscle cramps (including the abdomen), sweating, swollen eyelids, nausea, vomiting, headache, and

The female black widow has a reddish hourglass on its abdomen.

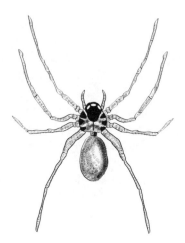

The brown recluse has a violin-shaped patch on its head and midregion.

hypertension. To treat, clean the site well with soap and water. Apply a cool compress over the bite location, and keep the affected limb elevated to about heart level. Persons younger than sixteen and older than sixty, especially those with a heart condition, may require hospitalization. Healthy people recover rapidly in two to five days.

Brown recluse
The brown recluse spider is ¼ to ½ inch long, has a yellowish to brown color, and supports a distinct violin-shaped patch on its head and midregion. Its bite causes a long-lasting sore that involves tissue death and takes months to heal. In some instances, its bite can become life threatening. The bite initially causes mild stinging or burning and is quickly followed by ulcerative necrosis that develops within several hours to weeks. The initial sore is often red, edematous, or blanched, and a blue-gray halo often develops around the puncture. As time passes, the lesion may evolve into ashen pustules or fluid-filled lesions surrounded by red, patchy skin. After several days, the tissue begins to die. Other symptoms include fever, weakness, rash, muscle and joint pain, vomiting, and diarrhea. To treat, clean the site with soap and water, immobilize the site, and apply a local compress. Use a pain medication if needed. The bite site ultimately needs

rapid debridement. Transport the victim to a medical facility as soon as possible.

Tarantulas

Tarantulas have hairy bodies and legs and come in a wide range of colors, from a soft tan through reddish brown to dark brown or black. The desert tarantula can grow to be 2 to 3 inches long and is common to the Sonoran, Chihuahuan, and Mojave Deserts of the U.S. Southwest. When confronted, a tarantula will rub its hind legs over its body, brushing off irritating hairs onto its enemy. Skin exposed to this hair is prone to an itching rash. A tarantula bite to humans is rare, and even if venom is injected, it rarely causes more than slight swelling, numbness, and itching. To treat a tarantula bite, clean the site with soap and water, and protect against infection. Treat skin exposures to tarantula hairs by removing the hairs with tape.

SCORPIONS

Scorpions are among the best-adapted creatures to desert climates. The scorpion has a flat, narrow body, two lobsterlike claws, eight legs, and a segmented abdominal tail. Its upward and forward curved tail has a

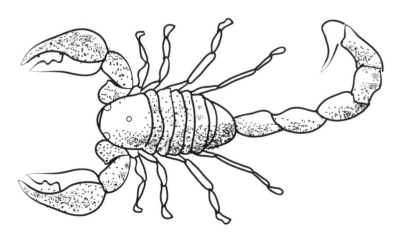

The scorpion has a venomous stinger at the end of its tail.

venomous stinger supplied by a pair of poison glands. Most scorpions are tan to brown in color and range from 1 to 8 inches in length. Their stings are generally painful but not fatal to humans. Other symptoms may include swelling at the site of the sting, numbness, muscle twitching, difficulties in breathing, and convulsions. Death is rare. There are a few species, some twenty worldwide, whose venom is potentially fatal, but survival rates are generally high. A scorpion's poison is neurotoxic, and treatment should follow that of a neurotoxic snakebite.

SURVIVAL TIPS

LEAVE DESERT CREATURES ALONE
Most desert creatures would prefer to be left alone and will bite or sting only when provoked. Thus, the best way to avoid harm is to avoid disturbing them.

Survival and First-Aid Kits

Obtaining prior knowledge and skill in the basic elements of survival is the key to ensuring a safe wilderness experience. In addition, proper preparation will prevent poor performance in almost any survival setting and will ultimately reduce the amount of stress you might experience. Central to your preparation are adequate survival and first-aid kits. These kits will play an instrumental role in how you meet your needs.

SURVIVAL KIT

Many factors will influence what you take, including cost, storage space, the size of your team, and where you're traveling. When putting together your survival kit, consider the five survival essentials.
1. Personal Protection
   Clothing. A dry change of clothes.
   Shelter. Tarps, blankets, space blankets.
   Fire. Metal match, lighter, matches, tinder.
2. Signaling. Cell phone, signal mirror, flares and smoke devices, fresh batteries.
3. Sustenance
   Water. Water purification tablets, water purifier, water storage containers.
   Food. Freeze-dried foods, vitamins, fishing tackle, snare line.
4. Travel. Magnetic compass, map of area, watch.
5. Health
   Psychological stress. Family photo, religious material, something to read.

Traumatic injuries. Carry an adequate first-aid kit.
Environment injuries. Sunglasses, sunscreen.

Other items. Flashlight, cup, plastic bags, fixed-blade knife, pocketknife, knife sharpener, nonlubricated condoms for holding water, scissors, tweezers, routine medications, moleskin, lip balm, sponges, light sticks, nylon string, long needlenose pliers, sewing kit, aluminum foil.

FIRST-AID KIT

- Aspirin
- Snakebite kit
- Water purification tablets
- Scissors
- Sunscreen
- Routine medications
- Matches
- Band aids
- Emergency blanket
- Bee sting kit
- Antihistamine
- Tincture of benzoin
- Roll of gauze
- Medical tape
- Triangular bandage
- Tweezers
- Moleskin
- Various dressings
- Lip balm
- Soap
- Antibiotic ointment

Cordage, Knots, and Lashes

IMPROVISED CORDAGE

Since line is key to holding improvised items together, you may at times need to improvise some cordage. Improvised cordage can be made from various materials such as cattail or yucca leaves, grasses, dried inner bark from some trees, animal products such as rawhide and sinew, or various man-made products such as parachute line or twine. The best rope-making materials have four basic characteristics: The fibers need to be long enough for ease of work, strong enough to pull on without breaking, pliable enough to tie a knot in them without breaking, and have a grip that allows them to bite into one another when twisted together. Any material that meets these criteria should work.

The first step to making cordage is to create long single strands of your selected material. Twist the material between your thigh and palm, adding additional fibers to its free end to create one long, continuous cord. This spun cord is then used in making two-strand, three-strand, or four-strand braid.

TWO-STRAND BRAID

Two-strand braid is an excellent all-around line that can be used for many tasks. If a lot of weight will be applied to the line, however, a four-strand braid would be a better option. Follow these steps to make a two-strand braid:

1. Grasp a piece of spun cord between the thumb and forefinger of your left hand, with two-thirds of its length on one side and a third on the other.

good braid

bad braid

A properly spun two-strand braid has even tension throughout.

2. With the thumb and pointer finger of your right hand, grasp the strand that is farthest away from you. Twist it clockwise until tight, and then move it counterclockwise over the other strand. It is now the closer of the two.
3. Twist the second strand clockwise until tight, and then move it counterclockwise over the first strand.
4. Repeat this process until done.
5. Splicing will need to be done as you go, and this is the reason for the two-thirds and one-third split. If you were to splice both lines at the same location, it would cause a significant compromise at that point. Splicing is simply adding line to one side. Make sure to have plenty of overlap between the preceding line and the new one, and use line of similar diameter.
6. To prevent the line from unraveling, finish the free end with an overhand knot.

If you are in a hurry and need a piece of short line right away, there is a quicker alternative:
1. Using spun cord, grab one end between the thumb and forefinger of your left hand, and roll it in one direction on the thigh with your right palm.

2. Repeat this process until the whole line is done and is tight.
3. Still holding the line at one end with your left hand, grasp the other end with your right.
4. Place the middle between your teeth, move your hands together, and tightly hold both ends in one hand.
5. Release the line from your mouth. The tension created by rolling the line on your leg will cause the two strands to spin together.

THREE-STRAND BRAID

A three-strand braid is ideal for making straps and belts. Follow these steps to make a three-strand braid:

1. Tie the three lines together at one end, and lay them out so that they are side by side.
2. Pass the right-side strand over the middle strand.
3. Pass the left-side strand over the new middle strand.

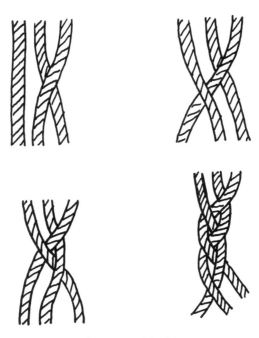

Three-strand braid

4. Repeat this process, alternating from side to side—right over middle, left over middle—until done.
5. To prevent the line from unraveling, tie the end.

FOUR-STRAND BRAID

A four-strand braid is ideal for use as a rope. It provides the strength and shape desired and is far superior for this purpose than either the two-strand or the three-strand braid. To make a four-strand braid, follow these steps:
1. Tie the four lines together at one end, and lay them out so that they are side by side.
2. Pass the right-hand strand over the strand immediately to its left.
3. Pass the left-hand strand under the strand directly to its right and over the original right-hand strand.
4. Repeat this process, alternating from side to side—right strand over strand immediately to its left, left strand under the strand immediately to its right and over the next one.
5. Splice in new material as needed.

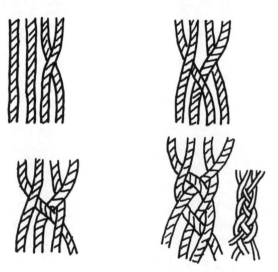

Four-strand braid

KNOTS

Several knots can be tied using man-made or improvised line.

SQUARE KNOT

The square knot connects two ropes of equal diameter.

DOUBLE SHEET BEND

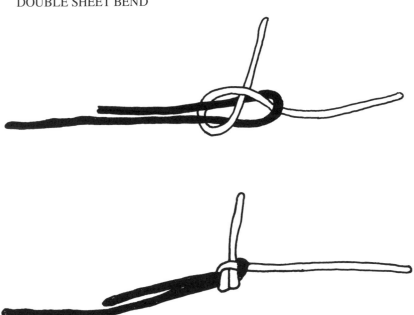

The double sheet bend connects two ropes of different diameters.

IMPROVED CLINCH KNOT

The improved clinch knot is used to attach a hook to line.

OVERHAND FIXED LOOP

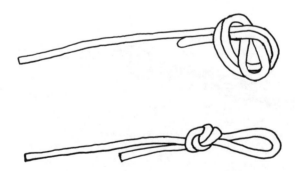

The overhand fixed loop has multiple uses in a survival setting.

BOWLINE

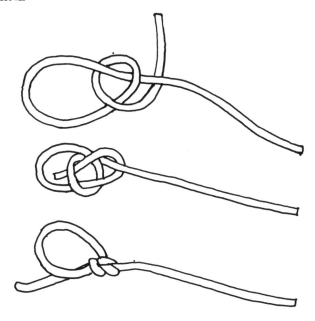

The bowline is much easier to untie after you use it than the overhand fixed loop.

DOUBLE HALF HITCH

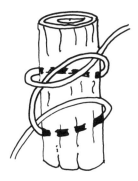

The double half hitch secures a line to a stationary object.

LASHES

SQUARE LASH

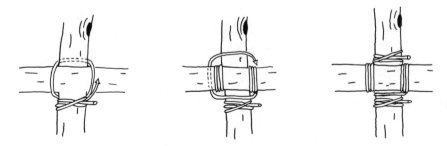

The square lash secures two perpendicular poles together.

SHEAR LASH

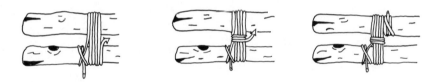

The shear lash attaches several parallel poles together.

Trip Planning

A trip plan should be completed and left with a reliable friend prior to your departure. If you fail to return or do not check in as scheduled, your friend should notify the sheriff or rescue organization. Plans are not filed with the sheriff's office, so unless your friend notifies the sheriff, a rescue mission will not be started. To avoid unnecessary searches, notify your friend of any delays and contact him or her immediately upon returning. Following is a sample trip plan. Modify it as appropriate for your personal use.

Person Reported Overdue

Name _____   Phone _____

Address _____   Age _____

_____   Medical conditions _____

_____   _____

Survival Equipment

Cell phone and number _____

Flares _____   Signal mirror _____

Smoke signal _____   Other signals _____

Trip Expectations

Depart from _____

Departure date _____   Time _____

Surviving the Desert

Going to _____

Arrival date _____ Time _____

Expected camping sites and water sources along the way:

	Date	Site (latitude and longitude)	Water source
1.			
2.			
3.			
4.			
5.			

If person has not arrived/returned by:

Date _____ Time _____

Call the sheriff or local authority at the following number

_____

Vehicle Description

License no. _____ Make _____

Model _____ Color _____

Where is vehicle parked? _____

_____

Other Persons on Trip

Name	Age	Phone	Medical conditions

Additional Information

Index

acute mountain sickness, 219
A-frame, 54–55
altitude illnesses, 218–220
anchors, 16–17
animals, 9, 57
   bites, 222–230
   handheld weapons for catching, 145–149
   as indicators of water, 100–101
   preparing game, 161–164
   snaring and trapping, 149–161
ants, 226

backpacks, 10–12, 199–200
bacteria, 112
bees or wasps, 225
birds
   as food, 143–145
   as indicators of water, 100
bivouac bags, 15, 45
blankets, emergency all-weather, 15–16
blankets, fleece or quilted, 19
blisters, 40, 214
blood, drinking, 99–100
bola, 146–147
bottle purifiers, 115
bow-and-drill technique, 64–66

cacti
   eating prickly pear, 124
   water from, 104
CamelBak, 12–14
campsites
   See also shelters
   locating, 44–45
caves, 55–56
cellular phones, 94
centipedes, 226

chlorine bleach, purifying water with, 113–114
climate, 9
clothing
   boots, 39–40
   gloves, 40
   headgear, 40–41
   how to wear and care for, 34–36
   layering, importance of, 31–32, 35 36
   natural materials for, 32
   parka and rain pants, 38–39
   shirts and pants, 36–38
   socks, 40
   synthetic materials for, 32–34
   waterproof, 33–34
compass, using a. *See* navigating
cooking methods, 164–167
cooking pots, 25
cordage, 233–236
crustaceans, 133–134

deserts
   characteristics of, 8–9
   types of, 5–8

edibility test, universal, 118–120
edible plants. *See* plants, edible
eye protection, 42

fire plow technique, 73–74
fires
   banking, 85
   bundle, 85
   fuel, 80–82
   kindling, 80
   lighting tinder from friction, 74
   maintaining, 84–85

man-made heat sources, 58–59, 60–63
natural friction-based sources, 63–75
oxygen requirements, 75–76
signaling using, 95
tinder, 76–79
fires, building
  bow-and-drill technique, 64–66
  fire plow technique, 73–74
  hand drill technique, 66–69
  location for, 59
  platform and brace, 75–76
  pump drill technique, 69–73
  steps for, 82–83
firewall, construction of, 83–84
first-aid kits, 232
fish
  gill net, 139–140
  hooks, improvised, 136–137
  injuries from hooks, 215
  lines, improvised, 137–139
  preparing, 142–143
  scoop net, 140
  spear/spearing, 140–141
  trap, 142
flares, 62–63, 87–88
flints, artificial and steel, 60–62
floods, flash, 204
food
  backpacking with, 117–118
  birds, 143–145
  cooking methods, 164–167
  crustaceans, 133–134
  fish, 136–143
  insects, 132–133
  mammals, 145–164
  mollusks, 134
  plants, edible, 118–131
  preserving, 167–169
  reptiles, 135–136
  storage of, 170
frostbite, 217
fuel, natural, 80–82
fuel tablets, solid compressed, 59, 77

gear
  anchors, 16–17
  backpacks, 10–12, 199–200
  bivouac bags, 15, 45
  blankets, emergency all-weather, 15–16
  blankets, fleece or quilted, 19
  CamelBak, 12–14
  cooking pots, 25

  headlamps, 25
  knives, 20–23
  poncho or tarp, 15
  saws, 23–24
  sleeping bags, 17–19
  sleeping pads, 20
  stoves, backpacking, 24–25, 58
  tents, 14–15, 45
  tips, 25–26
Global Positioning System (GPS), 198

hand drill technique, 66–69
headgear, 40–41
headlamps, 25
heat exhaustion, 216
heat gain and loss, 28–31
heat rash, 215
heatstroke, 216
helicopter rescues, 97
high-altitude cerebral edema (HACE), 220
high-altitude pulmonary edema (HAPE), 219–220
hobo shelter, 56–57
hydrated, staying, 207
hyponatremia, 216
hypothermia, 217

injuries/illnesses
  altitude, 218–220
  animal bites, 222–230
  bleeding, 209–211
  blisters, 40, 214
  bowel disturbances, 220–221
  burns, 213–214
  cold, 216–217
  fractures, 212–213
  heat, 215–216
  immersion, 217–218
  shock, 211
  sprains and strains, 213
  from thorns, splinters, and spines, 214–215
  traumatic, 208–215
insects
  bites/stings, 225–230
  edible, 132–133
  as indicators of water, 100
iodine, purifying water with, 114

kick-stepping, 200, 201
kindling, 80

knives
   cleaning, 22–23
   maintenance of, 21–22
   types of, 20–21
knots, 237–239

lashings, 48, 240
lean-to, 55
lighters, 60
lizards, 135, 224

maps, using. *See* navigation
matches, 60–62
mirages, 204
mirrors, signaling with, 89–91, 96–97
mollusks, 134
mosquitoes, 226
muscle cramps, 215–216

navigating
   checklist, 189
   compass, nomenclature, 177–179
   field bearing, establishing a, 185–188
   Global Positioning System, 198
   kamal device, 193–194
   location, determining, 180–185
   maps, nomenclature, 171–177
   maps, orienting, 181–182
   stars, using for, 194–197
   stick and shadow, 190–191, 197
   sun, using for, 190–194
   triangulating to determine location, 185
night, traveling at, 198

plants
   obtaining water from, 104–107
   survival strategies used by, 123–124
plants, edible
   acorns, 129–130
   amaranth, desert, 128
   berry rule, 121
   cattails, 130–131
   century plant, 125–126
   date palm, 128–129
   grasses, 124
   parts that are edible, 121–123
   prickly pear cacti, 124
   sotol, 124–125
   universal edibility test, 118–120
   yucca, 126–127
plunge-stepping, 200, 201

ponchos, 15
protozoans, 112
pump drill technique, 69–73
purifying water, 113–115

roofing
   birch and elm bark shingles, 53
   grass, 51–52
   mat, 52–53
   wood shingles, 54

salt water, 99
sandstorms, 204
saws, 23–24
scorpions, 229–230
shelters
   A-frame, 54–55
   cave, 55–56
   emergency tarp, 45–46
   hobo, 56–57
   insulating, 57
   lashings, 48
   lean-to, 55
   locating, 44–45
   shade, 46, 47
   tents or bivouac bag, 14–15, 45
   wickiup, 49–50
   wigwam, 50–54
signaling
   aerial flares, 87
   cellular phones, 94
   fire, 95
   ground-to-air pattern, 92–94, 97
   handheld red flare, 88
   improvised, 94–97
   kites, 91–92
   mirror with sighting hole, 89–91, 96–97
   parachute flares, 87–88
   rules of, 86–87
   smoke signals, generating, 95
   smoke signals, orange, 88–89
   strobe lights, 92
   tips, 97
   whistles, 92
skin protection, 42–43
sleeping bags and pads, 17–19, 20
slingshots, 147–148
smoke signals, 88–89, 95
snakebites, 222–224
snakes, 135–136

Index 247

snares
   Apache foot, 159–161
   figure-four mangle, 154–157
   Paiute deadfall mangle, 157–158
   simple loop, 149–152
   squirrel pole, 152
   twitch-up strangle, 152–154
Spark-Lite, 61–62
spear/spearing fish/animals, 140–141, 146
spiders, 227–229
Sterno, 58–59
stoves, backpacking, 24–25, 58
stress, 221
strobe lights, 92
sun blindness, 218
sunblock, 42–43
sunburn, 215
sun protection factor (SPF), 27–28
survival, steps for, 1–3
survival kit, 231–232

tarps, 15, 45–46
temperatures, extreme, 204
tents, 14–15, 45
terrain issues, 202–204
ticks, 226
tinder, man-made
   compressed tinder tabs, 77
   fuel tablets, solid compressed, 59, 77
   petroleum-based, 76
tinder, natural
   bark, 77
   grass, ferns, leaves, and lichen, 79
   wood scrapings, 79
Tinder-Quik tab, 76
transpiration bag, 106–107
traps
   See also snares
   box, 159
   for fish, 142

travel
   by car, 204–205
   hazards, 204
   techniques, 200–202
   tips, 206
   traversing, 200, 202

ultraviolet (UV) rays, 14, 27, 28, 42–43
ultraviolet protection factor (UPV), 27–28
urine, drinking, 99

vegetation, 9
vegetation bag, 105–106
viruses, 112

water
   amounts needed by the body, 116
   basin lakes, 103
   boiling, 113
   chemical treatment of, 113–114
   filtration, 110–112
   groundwater, 103–104
   importance of, 98–99
   impurities, 112
   indications of, 100–101
   myths about, 99–100
   from plants, 104–107
   purifying, 113–115
   rain, 104
   in rivers and creeks, 102–103
   in rock depressions, 103
   springs, 103
   surface, 101–103
water sources
   man-made, 108–110
   natural, 101–107
whistles, 92
wickiup, 49–50
wigwam, 50–54